PALESTINE

LAND OF PROMISE

Thou shalt inherit the holy earth as a faithful steward conserving its resources and productivity from generation to generation. Thou shalt safeguard thy fields from soil erosion, thy living waters from drying up, thy forests from desolation, and protect thy hills from overgrazing by the herds, that thy descendants may have abundance forever. If any shall fail in this stewardship of the land, thy fruitful fields shall become sterile stony ground or wasting gullies, and thy descendants shall decrease and live in poverty or perish from off the face of the earth.

"The Eleventh Commandment" written and broadcast over the radio by Dr. Lowdermilk in Jerusalem during June 1939 was dedicated to the Palestinian Jewish villages whose good stewardship of the earth inspired this idea.

PALESTINE
LAND OF PROMISE

BY

WALTER CLAY LOWDERMILK

GREENWOOD PRESS, PUBLISHERS
NEW YORK   1968

PALESTINE, LAND OF PROMISE

Copyright, 1944, by Walter C. Lowdermilk. Printed in the United States of America. All rights in this book are reserved. No part of the book may be reproduced in any manner whatsoever without written permission except in the case of brief quotations embodied in critical articles and reviews. For information address Harper & Brothers

REVISED EDITION

C-W

Reprinted with the permission of
Harper & Row, Publishers

First Greenwood reprinting, 1968

LIBRARY OF CONGRESS catalogue card number: 68-23308

PRINTED IN THE UNITED STATES OF AMERICA

The author wishes to make clear that this book was written from the point of view of the Land Conservationist whose life work has been to study the relation of peoples to their lands. The opinions expressed here are personal and unofficial. They do not necessarily represent the point of view of the U. S. Soil Conservation Service of which the author is assistant chief, or of any other government department.

CONTENTS

· · · · · · · · · · · · · ·

ILLUSTRATIONS

. . . .

*These illustrations will be found in a group
following page 116*

ILLUSTRATIONS

PALESTINE

LAND OF PROMISE

CHAPTER I

INTRODUCTION TO PALESTINE

. . . .

DURING recent generations, archaeologists have made a very substantial contribution to our knowledge of human history. By learning to read the records of old countries carved in stone or written on clay tablets, parchment and papyrus, they have been able to reveal to us the state of civilization at different periods of history. By excavating old and sometimes prehistoric sites of human settlement, they have traced the technological progress of humanity from its earliest beginnings.

There are, however, other records written by farmers and shepherds, empires and civilizations into their lands. Some tillers of soil were "good stewards" of the earth loaned to them by their Creator, while others let this primary source of all wealth fall into utter neglect. By good stewardship of the land some peoples have been able to preserve their fields through thousands of years of use so that they still yield abundant crops. Other fields are sorry commentaries on man's ruthless exploitation of the good earth. By neglect, ignorance and suicidal agriculture, peoples have bequeathed to their descendants "man-made deserts" of sterile, rocky and gullied lands.

Few historians understand the indelible marks left on the land by changing civilizations. For those, however, who have learned to understand their simple language, these records written by man into the land are extremely revealing.

For the past twenty years it has been my profession to read the record of the land written by changing generations. These studies give us a more profound knowledge of the human past, and at the same time serve as an imperative warning for the future. In the course of my work I have had opportunity to read the impact of past civilizations on the lands of China, Korea, Japan, Europe, North Africa, South Africa, the Near East, as well as our own beloved country.

It was in 1939 that my dream of many years came true; I was enabled to study the land record of the Near East. Palestine was of special interest to me because the Bible presents the most authentic and longest written record of any nation except China. Indeed, the peoples of these sacred lands of the Near East are responsible for much that makes the religious, political and educational institutions of the Western Hemisphere full of meaning for us. Moreover, we are indebted to the prehistoric farmers of the Near East for our basic principles of agriculture and for the seed-grains of wheat and barley that are the basis of bread, "the staff of life."

My study of the land record of the Near East was made possible by the United States Department of Agriculture which in 1938 sent me as a soil conservationist to make a survey of the use of land in these old countries in the interests of land conservation in the United States. Preceding our studies in Palestine were several months of investigation of the land records of Europe. We crossed afterwards from Italy to Sicily and then by boat to North Africa. We did all our land traveling by automobile, which nowadays is the best means to study rural areas of special interest. From Tunisia we traveled westward along the coast to Morocco and then turned eastward and slowly worked our way back through Algeria, Tunisia and 1200 miles along the Mussolini Highway in Tripoli and Libya and through the western Egyptian desert to the old land of the Pharaohs. We visited sites of special interest for a soil conservationist and almost everywhere we saw repugnant evidences of deadly soil erosion superseding the results of skilled land use during previous centuries. Egypt, which has preserved most of its ancient irrigation canals and added new irrigation dams and canals, is a notable exception.

Our further journey to Palestine met with great difficulties. The country was at that time disturbed by a prolonged period of Arab riots. No automobile for six months had crossed the Sinai desert dividing Palestine from Egypt, and travel there was extremely dangerous. It required several days of pleading and insistence before the Egyptian authorities granted us permission to cross Sinai in our car—and strictly at our own risk.

In February, 1939, we, like the Children of Israel, left the land of Egypt before daylight, well provided with provisions for ourselves and the car. We crossed the southern part of the Land of Goshen, which Joseph had given to his brothers because it was reputed to be the best grazing land in all Egypt. Now it is a desolate, bleak and barren region.

Finally, we entered the Sinai Desert, where the Israelites and their flocks and herds wandered for forty years. Today, this land maintains only a sparse population of nomads. Thousands of goat paths, like festoons of dismal draperies, twine in and out around the sides of barren hills. We saw gullies, cutting headward back into denuded slopes, revealing long abuse and extensive overgrazing. Passing this tawny arid region, with its sandy wadis and occasional thorny acacia, we were glad that modern science enabled us to traverse in two days by automobile a region where the Israelites wandered for forty years.

The first impression made on us by present-day Palestine was a depressing one. The Negeb, that sparsely populated semi-desert region of southern Palestine where we entered first, showed definite traces of a long period of Arab riots, which had hardly subsided at that time. Telegraph wires were dangling, border stations were wrecked and bridges were in ruins. Suspicious-looking Arabs stood around their tents, staring wonderingly but ominously at our lone automobile as it suddenly appeared on a road over which no one had traveled for months. We were careful to travel faster than a racer camel could run, in order that news of our approach should not be passed on to terrorist groups along the road, giving them time to start an attack on us.

Our reception by officials of the Mandate Government was very cordial. Every facility was placed at our disposal for studies in Palestine and Trans-Jordan. In districts menaced by Arab terrorism the authorities supplied an armored car to escort us, and Tommies to protect us while I took pictures and soil samples or made studies on foot. Also an airplane was put at our disposal with Captain P. L. O. Guy in command, for an additional survey of Palestine from the air.

In this small sacred land, we traveled more than 2300 miles by automobile and an additional 1000 miles in Trans-Jordan, besides making airplane surveys. During more than three months of intensive field studies, we were assisted by leading

[4]

officials of the Departments of Agriculture, Horticulture and Forestry, by soil scientists from the Hebrew University and by archaeologists, who accompanied us to all places of special interest in Palestine and Trans-Jordan.

During our stay in Palestine, the elements also co-operated in our work. A fine demonstration of the "latter rains" let us see how erosion had been carrying away the soils as a result of the neglect and breakdown of terraced agriculture. We saw drainage channels running full of brown silt-laden gully washers, cutting their banks and joining with water from other drainages to make a storm flood that roared down the main valleys. Here before our eyes the remarkable red-earth soil of Palestine was being ripped from the slopes and swept down into the coastal plain and carried out to sea, where it turned the blue of the Mediterranean to a dirty brown as far as the eye could see. We could well understand how during many centuries this type of erosion has wasted the neglected lands. It is estimated that over three feet of soil has been swept from the uplands of Palestine since the breakdown of terrace agriculture.

Along with the records of decay in the Holy Land we found a thoroughgoing effort to restore the ancient fertility of the long-neglected soil. This effort is the most remarkable we have seen while studying land use in twenty-four countries. It is being made by Jewish settlers who fled to Palestine from the hatreds and persecutions of Europe. We were astonished to find about three hundred colonies defying great hardships and applying the principles of co-operation and soil conservation to the old Land of Israel. Amazed by this phenomenon we gave much time and attention to study of the methods and achievements of these colonies and the diffi-

[5]

culties they had to overcome. Here in one corner of the vast
Near East, thoroughgoing work is in progress to rebuild the
fertility of land instead of condemning it by neglect to further
destruction and decay.

The far-reaching works of reclamation now being conducted in Palestine not only provide an example to other
countries in their need to restore wasted lands but are vitally
important for hundreds of thousands for whom they mean the
only chance for security and human existence. The persecution and mass slaughter of Jews in Nazi-dominated Europe
have vastly increased the dynamic power of the building
activities in Palestine. Since 1933, Hitler's anti-Semitic drives
have deprived millions of Jews of their livelihood and driven
them from their homes and the lands of their birth. Many of
them dared all risks to reach the one "Land of Promise"
where they could find a haven of refuge. Here, their trained
minds and physical energies have been devoted to the establishment of a National Home for themselves and their descendants. The Jews, since 1917, have spent over half a billion
dollars on redeeming the land of this small country and
rehabilitating hundreds of thousands of refugees who are
only part of the millions needing a haven. The Jewish population of Palestine has grown from about 50,000 in 1918 to over
550,000 in 1943, representing over a third of the total population of about 1,600,000 of whom 900,000 are Moslem Arabs
and 125,000 are Christians.

The movement for establishing a Jewish homeland in
Palestine is one of the most remarkable records of a people's
struggle for national survival and self-expression. It began
about four thousand years ago in Ur of the Chaldees, when
Abraham, prompted by divine inspiration left the plains of

Mesopotamia to establish a new people on the Land of Canaan. It continued after several centuries, when the descendants of Abraham went to Egypt for food during a famine and were held there in bondage. Oppression by the Pharaohs fired them to throw off this slavery and again they directed their steps across the wilderness to their homeland. This urge came to life yet again when, after centuries of settlement in Palestine, the Jewish people were driven into exile by the conquests of Assyria and Babylon. On the rivers of Babylon the exiled Jews continued to dream of returning to their devastated National Home, and their leaders, Ezra and Nehemiah, finally went back into Palestine to rebuild the walls of Jerusalem.

After centuries of prosperity, the Jewish people again were caught in the conquests of an aggressor—this time it was Rome—and after heroic resistance, they were driven from their Sacred Land. Since then Jewish craving for Palestine has been the main inspiration of this distressed people. This dream of the restoration of Zion has never died out in the two thousand years of the Diaspora. The religious faith of the Jews has been permeated by a fervent belief in a prophetic resurrection of the Jewish nation in the land of their forefathers. They have been a minority in other lands and have borne the brunt of hatreds and persecutions to which minorities are so often subject. When the industrial revolution came to eastern and central Europe and made life intolerable for small Jewish tradesmen, craftsmen and professional men, a new inspired leader, Dr. Theodor Herzl, rekindled with prophetic vision the old dream of a homeland in Palestine. Hardships and anti-Semitism in Europe drove these persecuted people in ever-growing numbers to seek a better way of life in Palestine. During the First World War the justice of their historic aspiration was recognized by the British Balfour Declaration of November 2, 1917, reading:

His Majesty's Government view with favor the establishment in Palestine of a National Home for the Jewish people, and will use their best endeavors to facilitate the achievement of this object, it being clearly understood that nothing shall be done which may prejudice the civil and religious rights of existing non-Jewish communities in Palestine, or the rights and political status enjoyed by the Jews in any other country.

Seconded in 1922 by a unanimous resolution of the United States Sixty-Seventh Congress and adhered to by the League of Nations, the Balfour Declaration became the charter rally- ing a world-wide Jewish crusade for the redemption of Pales- tine as a National Homeland. A great surge of colonization followed. But the necessity for a Jewish resettlement of Palestine took on its full urgency only with the rise of Nazism.

During my stay in Palestine in 1939, I witnessed a tragic by-product of the German advance into Czechoslovakia. In Palestine and Syria we were told of old cargo boats, filled with refugees from Nazi-dominated Central Europe, whose captains tried desperately to disembark this living cargo on the shores of Palestine. We saw some of these old and often unseaworthy boats, whose miserable passengers were not permitted to land anywhere because of the lack of formal visas. We saw those wretched ships floating about on a steaming sea in unbearable summer heat, with refugees packed in holds under intolerably inhuman conditions. The laws governing the transportation of animals for slaughter in the United States do not permit conditions like those which some of the intelligentsia of central Europe had to undergo in these old boats on the Mediterranean. The revolting slave ships of a century ago were better; for slaves had a sale value and their ships were sped to their destination without delay. But Jewish refugees were kept floating about upon a torrid sea, just out of sight of land, with a desperate hope that the captain, though risking confiscation of his ship, would attempt to discharge them illegally on the shores of Palestine. Unbelievable tragedies have taken place on the way to Palestine. I was told of ships that had set out laden with refugees, and after some months turned up empty again, with no trace of their human cargo. None of them have been heard from since by their relatives.

During our stay in Beirut an old cargo boat, loaded with 655 refugees from Czechoslovakia, was unloaded at the quarantine station for a few days. The ship was so overrun with rats that the passengers had to be removed to exterminate these vermin. After obtaining permission to interview the refugees, we found that they had been floating about for eleven weeks, packed into little wooden shelves built around the four cargo holds. The congestion, the ghastly unsanitary conditions and sufferings that these people had undergone aroused our highest admiration for their courage and fortitude. Their food was gone, and all of them had contracted scurvy from malnutrition. We were astonished to find that these former citizens of Czechoslovakia represented a very high level of European culture. Most of them spoke several languages, and many of them were able to tell us their story in English. Of the 655 refugees, forty-two were lawyers, forty were engineers, twenty-six were physicians and surgeons, in addition to women doctors, professional writers, gifted

musicians, pharmacists and nurses. Two had been staff officers of the Czech army before its dissolution by the Nazis,
sixty had been army officers, and two hundred, soldiers;
many others were skilled workers and craftsmen.

Without passports, without country, these useful and highly
cultured refugees presented one of the most tragic spectacles
of modern times. No ambassador, no consul spoke up for
them to demand the rights and privileges enjoyed by the
lowliest citizen of the smallest country. They were adrift
without a port of entry, without a representative who could
protect them. What a stigma upon our modern civilization!
What has become of our conception of the infinite value of
the individual?

Even after Hitler's defeat and the restoration of their civic
rights, many Jews who may survive this terrible period in
Europe will still need to flee the hatred of their erstwhile
persecutors. It will be the responsibility of the victorious
United Nations to find a place for them where they can re-
establish themselves on a constructive and secure basis.

Palestine with its unique possibilities of reclamation can
solve this age-old problem of persecution and homelessness.

To insure the full development of Palestine's remarkable
power and irrigation possibilities, we shall propose in the
course of this book a great reclamation project, tentatively
called the "Jordan Valley Authority" along the lines of our
own Tennessee Valley Authority. Before presenting the features and objectives of this great project we shall acquaint
the reader with the lands of the Near East, and especially
with the regions which are historically part of Palestine and
which logically fall into the scope of the proposed reclamation work. We shall examine the geographical setting of these

areas, the state of their population and the present achieve‹ ments of Jewish agriculture and industry in Palestine, as an indication of the country's ability to absorb a great refugee population. We shall describe how our reclamation project can enable this population not only to work out its own salvation but also to demonstrate how the Near East as a whole can be lifted to a much higher economic and cultural level.

CHAPTER II

A VENTURE IN CONSERVATION AND ITS
POST-WAR IMPORTANCE

. . . .

"The Earth is the Lord's and the fullness thereof,
the world and they that dwell therein."
—PSALMS, 24:1

FOR the past century the Near East has been a sore spot in world affairs. In fact, it has been called a "low pressure area of international storms." The backwardness and resulting weakness of its peoples invite manipulation and aggression by more powerful nations. The strategic position of the Near East at the crossroads between three great continents, and the existence of vast oil fields and certain other materials vital to modern industry and war, make it an attractive object for foreign rivalries and domination.

The helpless position of the Near East in present world politics is in full contradiction to its leading part in ancient times. At the dawn of history the peoples of this region started and developed to a high degree a civilization built on a remarkably intelligent use of land and water.

The Heritage of the Near East

Our civilization is deeply indebted to the sacred lands of the Near East. From them we inherited the origins of our present-day philosophy, science and religion. Jews, Christians and Moslems alike, look upon Palestine as their Holy Land; millions of pilgrims to its historic shrines have been stirred by its sacred and glamorous past. Out of Palestine has come the concept of the infinite value of the individual, of the common man, which is the basis of our democracy that we are now called on to fight for to the death.

Along with spiritual values, we have inherited from the Near East many basic inventions which laid the foundation of modern technology—the lever and screw from Egypt and the wheel from the ancient Sumerians of Mesopotamia. The latter likewise taught us to divide the day into hours and minutes, and the circle into degrees and subdivisions of 60 units.

For our calendar we are indebted to the Egyptians. Also from the Near East we inherited the grain foods of wheat and barley and the fundamentals of agriculture. Here in the alluvial plains, some genius of a farmer in the remote past hitched an ox to a hoe or other implement and thus invented the plow, and for the first time applied power to farming. With the discovery of the plow and the advantages of irrigation, a tiller of soil was able to produce more food than he needed for himself and family and released some of his fellow villagers for other tasks. Thus began the division of labor which made advances in civilization possible. As more and more persons were released from the task of producing food, and no other ways were found to put them to work, unemployment must have become a problem, resulting in social unrest. As we gazed on the brooding Sphinx and the great pyramids, the largest of which alone took 100,000 men twenty years to build, I had a suspicion that these gigantic pyramids of Egypt were the first great W.P.A. projects.

Whenever, due to historic changes, the people of the East neglected the conservation work of their ancestors, a general decay resulted. Particularly, during most of the past 1200 years, lands of the Near East have been gradually wasting away; its cities and works have fallen into neglect and ruin; its peoples also slipped backward into a state of utter decline.

In the midst of general decadence in the Near East, there is hope in Palestine; hope made real by the Jewish colonies which are showing the most remarkable devotion to reclamation of land that I have seen in any country of the New or Old World. The result of their efforts thus far is an inspiration and splendid achievement. Unknown to themselves, these colonies have laid the foundation for a Greater Palestine and have shown the way for the resurrection of the Near East as a whole.

The Jews have proved themselves capable of this herculean task of reclaiming the long neglected Holy Land. Since 1882 they have been struggling against great odds to colonize in Palestine and to restore it to a high level of fertility and civilization. With religious zeal and sacrificial martyrdom they have flung themselves into this cause of reconstruction and redemption of wasted and depleted lands of Palestine. The way in which these changes are being wrought in the land is one of the most remarkable phenomena of our day. The Rev. Norman McLean, chaplain to King George VI, recently stated "There is no experiment in human uplift, now to be seen on the face of the earth that can compare to the work of the Zionists in Palestine. If I were a Jew, I would deem it the highest honor life can hold to have a part in a work so noble." This work in Palestine is also a great inspiration to land conservationists everywhere.

In some ways, colonization in America was like the colonization of Palestine—there were hardships and dangers in both cases. But in the case of Palestine, it had to be done on old and impoverished soil rather than on new and bountiful land. Palestine has become a gigantic experimental area, full of contrast between land-use under primitive and backward conditions, and the most advanced methods of treating land and water resources. The remarkable enterprise of Palestine colonization brings to the fore in their simplicity the fundamental factors of man's relation to the earth. For our complex industrial civilization masks and obscures the processes directing and governing the adjustment of populations to the resources of the earth.

Land Use—A Post-War Problem

The problem of soil and water conservation, so decisive in the life of ancient civilizations, will acquire an additional importance after the end of the present global war. This war will not be ended by the mere cessation of hostilities. Great world problems will have to be solved, and deep wounds will have to be healed by many years of united effort. Otherwise the end of this war will only prove to be a pause before a new and more horrible holocaust.

To grasp the post-war problems facing the United Nations after the end of this war we must understand that the world is going through a major break-up of exploitive economy based on rapid occupation of new lands and exploitation of backward peoples. This upheaval in the civilization of our time is of even greater magnitude and significance than the break-up of the Roman Empire 1500 years ago. It is essentially a blind effort on the part of mankind to read-just itself to the land and to gain access to its products for the common man.

This disintegration of our pre-war civilization has been hastened by our scientific age when power applied to machines has greatly increased the effectiveness and efficiency of those who know how to use them. It sharpens and makes all the more glaring the contrasts between peoples who have industrialized their countries, and those who are still in a state of agrarian subsistence. Countries advanced in industrial techniques, have sought out raw materials needed by their industry in every corner of the world. These raw materials include not only foods for human consumption, but vital and rare materials required to feed machines in an industrialized society. They were partly obtained through commerce with other powerful peoples, but they were also gained to a great extent by conquest and exploitation of backward peoples who were ignorant of the use of essential raw materials within their own borders.

By taking advantage of the low standards of living of backward peoples, nations more advanced in technology and finance have made many toilers of backward countries work for their own industrial economy. In this way backward peoples are exploited and made to give to more advanced peoples enviable luxuries and leisure. They are forced to work much harder for bare necessities of life than the well-favored in industrialized countries have to work for luxuries. When conscious of their exploited weakness, these backward peoples rebel in spirit before they are able to rebel in arms. This world-wide war gives them an urge to free themselves from the heavy hand of exploitation. The century of the common man is at hand.

Whatever may be the justification of past policies, advanced peoples find it more than ever to their interest from now on to provide ways for backward peoples to better themselves and to share in the good things of the earth. Failing to do this would result in another war of even more frightful proportions. For the human spirit will rise again, even when crushed to earth by most ruthless and brutal measures.

Exploitation or Conservation

The global war of today has developed into a gigantic contest between two groups of nations that uphold two opposing philosophies. The winning side may determine the general state of mankind for a thousand years to come.

The Nazi-Fascist-Japanese group fights to make the individual serve the State. This group seeks systematically to subjugate their own people in the interest of the State and, through conquest, to enslave other peoples and to seize their resources for exploitation by the self-styled "superior nations." Conquered peoples are being forced into vast sweatshops to work for their masters. All the gains of personal liberty and freedom have been blotted out by the super-State. The goal of the totalitarian nations is to continue the general policy of exploitation which nations have followed substantially since the beginning of history.

But through centuries of bloody struggles, another philosophy has been emerging. Its ideas were first proclaimed by

[17]

the Hebrew prophets in the Old Testament and were more completely stated in the Sermon on the Mount, delivered on the shores of the Sea of Galilee nearly two thousand years ago. Here the Great Teacher of mankind, speaking before a throng of common people, stressed the infinite value of each individual and his right to a fair deal based on freedom and equality. Here he proclaimed the "Golden Rule" of doing "unto others as you would have others do unto you." Scathingly he denounced injustice, oppression and the exploitation of the poor and weak by the rich and powerful. A more abundant life was proclaimed to be the heritage of all men, and spiritual values were recognized to be as vital in the development of individuals and nations as physical needs. "For man shall not live by bread alone."

It is to maintain our rights and liberties and to help other nations to regain their lost freedom that the United Nations fight today: it is to shape society in the interests of the individual, rather than of favored groups. Our goal is to make the State serve the individual more fully, to assist him in realizing his noblest dream and reaching his full stature. The upsurging of the spirit of freedom has given us the Atlantic Charter as the goal of our war effort today; it is a charter that must apply on the seven seas as well as on the Atlantic.

Whatever may have been the exploitive policy of members of the United Nations in the past, the only basis on which they can get together to fight the common enemy is that of freedom for the common man everywhere—not only within their own national borders. Only on this basis can each nation safeguard its own freedom. The time has come when society must be born again out of an economy of exploitation, into an economy of conservation in its widest sense, developing its

[18]

human resources and conserving the resources of the Good
Earth from generation to generation.

Food and Civilization

What are the vital relations of a people to its land?

It was in China, when studying ways to prevent the awful
scourges of widespread famines, that we were brought face
to face with man's primary needs. We learned that in the last
reckoning all things are purchased with food. The present
world-wide war only emphasizes the grimness of this awful
truth. Food is necessary to carry on the war. The aggressor
nations are rationing food to subjugate rebellious peoples in
occupied countries. Men will sell all—their liberty and more
—for food, if driven to this tragic choice.

Food, and not money, which is after all only a symbol, a
convenience in the exchange of goods and services, buys the
division of labor that makes possible advance in civilization.
It was not until the tillers of the soil produced more food
than they themselves had need of that their fellow villagers
were released for other tasks. This has been true whatever the
motives that prompted farmers to grow surplus food, whether
they were urged on by the whiplashes of slave drivers in
ancient Egypt or lured by the profit motive in our times.
Until food is available, the miner does not dig into the bowels
of the earth for minerals and ore, and mechanics do not
process ores and make the intricate machines of modern
technology. There is no substitute for food in the complex
division of labor in modern civilization.

Land is the silent partner of the tillers of the soil in the
growing of food. This partnership of land and farmer is the
rock foundation of our civilization; if either member of this
partnership weakens or fails, the whole structure of civiliza-

tion built upon it likewise weakens and fails. Nations rise or fall upon their food supply, and hence ultimately upon the condition of their land.

The relation of a people to its land is thus far more than economics. The land becomes an integral part of the nation, even as its peoples are. Aside from our social institutions and blood relations, our most enduring and influential link with future generations is the condition of the land as it is bequeathed to those who follow us. The survival of a people reflects the imperious mysteries of life itself. But if life is sacred, then the earth that nourishes it also becomes sacred. In the Laws of Moses it is stipulated that "The land shall not be sold forever, for 'the land is Mine,' saith Jehovah, 'and ye are strangers and sojourners with Me'." (Lev. 25:23)

The rich and powerful have for centuries exploited the farmer, and he in turn has exploited the land, often letting it become sterile and wasted. The "sins of the fathers," in wasteful and careless exploitation of their fields, are visited on their children, and on all future generations who toil in erosion-wasted and gullied fields. Neither a people nor a civilization may grow in strength, reach high standards of living and realize nobler hopes, unless the land base is saved and maintained in a productive condition. There can never be peace and stability in society or among the nations of the earth unless abundant and adequate food is made available. In the final reckoning, land can be separated neither from man's loftiest aspirations nor from his highest obligations.

The Eleventh Commandment

It was in Jerusalem, the Holy City of the "Promised Land," that I was amazed at the contrast between the present aspect of the surrounding country and the description of the Promised Land by Moses as he stood on Mount Nebo, in the region now separated as Trans-Jordan from the British Mandate of Palestine.

Behold, the Lord thy God giveth thee a good land, a land of water brooks and fountains that spring out of the valleys and depths, a land of wheat and barley, of vines, figs and pomegranates, of olive oil and honey, a land in which thou shalt eat bread without scarceness, thou shalt not lack anything in it.

—Deuteronomy 8:7-9

From such descriptions and from archaeological findings as well as from my own study of soils, present-day climate, the remnants of vegetation and other factors, we may conclude that the Land of Israel was capable of supporting and actually did support at least twice as many inhabitants as at present.

Backward native populations and political and social decay are the usual result when land is impoverished by erosion and neglect. Palestine is a classic example of such transformation. The contrast between Moses' description of Palestine and the state of the country when Jewish colonists first began their work in 1882 affords a graphic commentary on the misuse and neglect of resources in the Holy Land. The soils were eroded off the uplands to bedrock over fully one-half the hills; streams across the coastal plain were choked with erosional debris from the hills to form pestilential marshes infested with dreaded malaria; the fair cities and elaborate works of ancient times were left in doleful ruins.

The change for the worse is astonishing. Glancing at the country from the windows of a rapidly moving train or car, casual visitors of today consider as normal the rocky, semi-arid, run-down condition of much of Palestine. But those who are able to read the record that has been written in the land know that this state of decadence is not normal. They can tell from the ruins of terraces and of other ancient conservational works that the present desolation of Palestine is due to the plunder, exploitation and neglect of recent centuries.

It was while contemplating the tragedy of this land through the ages that I wondered what would have happened if Moses, when he was inspired in the wilderness to deliver to the Israelites the Ten Commandments governing their relationship to their Creator and to their fellow men, had been a farmer as well as a shepherd. Moses might then have foreseen man's neglect of the Promised Land. Would he not have been inspired to deliver still another Commandment establishing mankind's relation to Mother Earth?

In a radio broadcast from Jerusalem in June, 1939, I tried tentatively to formulate such an "Eleventh Commandment," and dedicated it to the Jewish colonies whose good stewardship in redeeming the damaged Holy Land was for me a source of great inspiration:

Thou shalt inherit the Holy Earth as a faithful steward, conserving its resources and productivity from generation to generation. Thou shalt safeguard thy fields from soil erosion, thy living waters from drying up, thy forests from desolation, and protect thy hills from overgrazing by the herds, that thy descendants may have abundance forever. If any shall fail in this stewardship of the land, thy fruitful fields shall become sterile stony ground or wasting gullies, and thy descendants shall decrease and live in poverty or perish from off the face of the earth.

[22]

A VENTURE IN CONSERVATION

Herein is expressed the over-all national responsibility for the land, a responsibility that may best be met by enlisting individual interest and initiative in accomplishing so great a task. This responsibility is as imperative as that of safeguarding the people's independence. For it is through the land that a government in the long run achieves social justice for its people, and provides for defense and general welfare.

Nations of today may be classified by the way they are fulfilling this responsibility. Until recently, our own United States stood on a low level in this respect, but during the past decade the movement for land conservation has lifted us to the highest place. We have made a good beginning, but only a beginning. There is still much to be done to save our lands from soil erosion—the insidious enemy which in the long run is more devastating than war.

In this period of human history when certain nations follow gangster leaders who hold out promises of new land as remedy and release from privations, the state of land use has become a matter of acute international interest. Exploitation of the soil, which leads to imperialism and war, has failed mankind. It is the principle of land conservation on which the hopes for a better future depend. Seemingly overpopulated countries can furnish their people with enough food, provided their resources are fully husbanded by the application of the techniques of conservation.

Reclamation in the Holy Land

After pondering for many years the problems of the adjustment of peoples to land resources, we approached the issues involved in the development of Palestine. In contrast with the doleful general state of lands in the Near East, we found in this small country, little larger than our state of Vermont and limited by the Mandate to 10,429 square miles, a series of profound experiments in the adjustment of a people to land resources. Driven by necessity and the intolerable sufferings of harried refugees, the Jewish settlements in Palestine were born. In them the white heat of the struggle for survival has burned out the dross of modern, luxury-loving life.

These experiments in colonizing a damaged and wasted land are planned to restore that land to high and sustained production and to build, on the foundation of this new agriculture, a social structure that will give to all members of the commonwealth, not only security and liberty, but cultural satisfaction as well. Here in the hallowed land of Palestine we see emerging before our eyes, on a scale easily comprehended, the principles that must underlie an enduring social order and civilization. Here have been worked out practical methods for achieving satisfactory standards of living for all members of the community alike.

In their Palestinian settlements Jews have shown the way for the reclamation and restoration of the decadent Near East. But they have not as yet even begun to exhaust the possibilities before them. For the type of work done in these settlements on a comprehensive scale will have to be applied where the development of all resources may be co-ordinated. The Tennessee Valley Authority has set the pattern whereby agriculture, power and manufacturing can be developed in a co-ordinated way in the highest interests of the people of a given area. This pattern should be applied to Palestine. The drainage area of the Jordan Valley, including Trans-Jordan, the Hauran and the maritime slopes of Palestine, has the physical features and resources that set the stage for a great reclamation project. This project will make possible the constructive solution of one of the most distressing problems of this war and of the post-war period, for it will provide room and work for millions of Jewish refugees now suffering persecutions in many lands of Europe.

CHAPTER III

GEOGRAPHICAL FACTS

. . . .

Geographical Background

PALESTINE is part of the Near East, sharing its topography, geological structure, climate, soils, flora and fauna and its human history.

The Near East includes Mesopotamia, which is now known as Iraq, Syria, Arabia, Lebanon, Palestine, Trans-Jordan, Sinai and Egypt. The two great alluvial plains of the Nile and the Euphrates cradled early man. As their population increased, it overflowed into the mountainous area of Syria and Palestine which separates these two fertile river valleys.

It was in these valleys of Mesopotamia and Egypt that agriculture first was developed. Under an arid climate, these fertile lands gave rise to farming by irrigation. They supported large populations whose struggles with each other and with grazing hordes out of the surrounding steppes and deserts made the dramatic history of the Sacred Lands. Indeed a tragic record has been written by this age-old conflict between the sown and the unsown, between farmer and shepherd, between a settled population and the wandering nomads of the desert.

[26]

GEOGRAPHICAL FACTS

The mountainous lands of Lebanon and Palestine, lying along the eastern shore of the Mediterranean, served as a highway between Mesopotamia and the Nile Valley. Their lofty hills condensed on their shoulders copious rain out of the moisture-laden storms that sweep across the land from the Mediterranean in the winter. These rains supported forests which supplied timber to the empires of Mesopotamia and Egypt, and provided abundant ground waters that issued forth as perennial springs in the valleys.

The rains also made it possible to farm on slopes without irrigation, thus introducing to the Near East a new and hazardous kind of land use leading to soil erosion. For the first time, tillers of the soil were faced with the most difficult problem of land use, that of establishing an enduring agriculture on sloping lands watered by heavy rains. In the area now called Lebanon, Palestine and Trans-Jordan the problem was solved by elaborate terracing. But these works required regular maintenance, for we have learned that terraced soils, if neglected, are more vulnerable to accelerated erosion than undisturbed soil.

Palestine has unusual interest for a geographer and soil conservationist. I found a fascinating record written on that land by the farmers and peoples who have occupied it for the past four thousand years. Briefly, this record tells how a land very much like Southern California and better favored by rains passed through five periods. First, it was occupied by hunters and shepherds. In the second period, farmers first cultivated the slopes of Palestine and induced destructive erosion such as has occurred in the United States in recent generations. During the third period, which we may call that of "terraced agriculture," farmers, with infinite patience and labor, carried out measures to conserve soil and rainfall, achieving the highest stage of prosperity for the largest population possible. The fourth period covers the breakdown of agriculture that took place between the seventh and twentieth centuries. In this period of decay, agriculture, which had previously achieved remarkable refinements in conservation, declined under exploitation by ignorant rulers and repeated invasions of nomads. The fifth period is now beginning—an era of restoration of the land to its possibilities as a source of human welfare and refuge.

The region in which we are specially interested includes the drainage of the Jordan River and the seaward slopes of Palestine. While arbitrarily cut into unnatural divisions by recent political boundaries dividing Mandate Palestine from Trans-Jordan and from small neighboring portions of Lebanon and Syria formerly a part of Palestine, it actually comprises a geographic unit of about 50,000 square miles in which the Jordan Valley is the most striking feature, vying with the Grand Canyon of the United States in its geological make-up and history.

Of this region, Palestine as it was designated under the British Mandate following the First World War, covers an area of 10,429 square miles. It borders on Syria in the north, and on Egypt in the south, while its western frontier is formed by the Mediterranean Sea. Its eastern frontier, as defined by the purely political division of the Mandate, is at the Jordan. On the other side of this historic river lies Trans-Jordan, a country of 37,740 square miles, ruled by an Arab Prince (Emir Abdullah) under British protection and supervision. But up to 1922 Trans-Jordan was considered a part of Palestine. In several respects (duties, railway system, responsibility for joint defense, and joint reports to the League of Nations) Palestine and Trans-Jordan are still treated as a unit. Trans-Jordan actually is a country "set up within the Palestine Mandate" but excluded from the provisions concerning the Jewish National Home.

The mandated area of Palestine is much less than the original area of historic Palestine which included Trans-Jordan up to a north-south line parallel to the present Hedjaz Railway. Historical Palestine likewise extended farther north into the frontier districts of the present area of Lebanon and Syria, and its southern area included the wilderness of Sinai.

Geological Foundation of the Land

Rock formations of limestone and sandstone form the geological structure of Palestine and have influenced the course of its human history to a remarkable extent. Other formations outcrop in limited areas and make up a lesser part of the visible skeleton of the land. Limestone formations give Palestine its characteristic features, its remarkable "red-earth" soils which are very fertile but subject to serious erosion on slopes. These formations provide underground drainage that bursts forth in a series of springs, some of which form ready-made rivers that played a decisive part in the early occupation of the land.

The physiographic region of which Palestine and Trans-Jordan are parts is a mountain fold of Mesozoic limestone that was uplifted in early Pliocene times along the eastern shore of the Mediterranean. We found evidence of this in the fossil oysters, clams and bivalves which we discovered in abundance over the rocky terraced fields high up on the Lebanon Mountains. The western margin of this fold was

[29]

washed by the Mediterranean Sea and the eastern margin was lost in the great Arabian desert. Ancient streams drained most of this land into the Mediterranean Sea. The pre-Ice Age Yarmuk River rose in the Hauran as it does today, but it flowed westward through what is now the Beisan and Esdraelon Plains into the Mediterranean Sea, reaching it near Haifa.

The ancient Yarmuk drainage also marks a division between northern and southern climatic zones that have prevailed in Palestine since Pliocene times. West-flowing drainages to the north of the Yarmuk received enough rainfall for streams to reach the sea (such as the Litany in Lebanon). But south of the Yarmuk drainage rainfall has never since the uplift supplied sufficient runoff waters for streams to cut their ways to the sea across the land barrier. This interesting fact written deep in the landscape indicates a remarkable uniformity of climate throughout this long period of geological time. Thus we know that the climate and rainfall have remained essentially the same in this region for several million years.

A most spectacular geological event took place before the Ice Age in this uplifted fold.[1] A series of north-south faults not far from the axis of this mountain fold let down a narrow strip of the earth's crust into the greatest inland depression of the earth, running parallel to the coast line from the Red Sea in the South to the Taurus Mountains of Turkey in the North. The most southerly part of this geological trench in the earth's crust let in waters of the Red Sea to form the Gulf of Aqaba; but the 1200-foot ridge in

[1] Bailey Willis, "Dead Sea Structure," *Bulletin of Geological Society of America*, Vol. 49, Apr., 1938, pp. 659-668.

the Araba dammed off the Red Sea from the Dead Sea and the Jordan depression. The Dead Sea depression was let down over 2600 feet below sea level at the lowest point; it was filled in by waters of the Dead Sea to a level still 1290 feet below that of the sea. To the north, the uplift of Mount Hermon raised the rift valley to its highest point near Baalbek, in the spectacular Bekaa gorge that cleft the Lebanon dome into two parallel ranges and the picturesque Lebanon and Anti-Lebanon Mountains famous of old for their sweet-scented forests. Northward the Orontes flows in the northern extension of the depression until it breaks out to the sea at the site of ancient Antioch.

The outstanding feature of the southern part of this faulted region is the Jordan Valley with its unique stream that drains the uplifted lands on the north, east and west. The source of the Jordan is two-headed, the springs of Nahr Leddan, and the Nahr Banias which issue forth out of the great limestone mountain mass of Mount Hermon at an elevation of three thousand feet above sea level. Two lesser streams join with the headwaters of the Jordan below Banias and flow into the papyrus marshes of Lake Huleh (Merom of the Bible), 230 feet above sea level.

After leaving the Huleh, the Jordan River soon falls below sea level, rushing down through a narrow channel out into a basaltic dam for a fall of 900 feet in the nine miles between the Huleh and the Sea of Galilee. This sea of sacred memory lies as a jewel set deep in a steep-walled depression in a basaltic rock that has weathered into fertile soils. In ancient times olive orchards and crops about the Sea of Galilee were famous for their luxurious growth and yield, but on our visit to Capernaum we passed along a dirt road lost in a thicket of thistles.

Southward from the Sea of Galilee the Jordan has cut for itself a narrow flood plain into unconsolidated beds that were deposited in the Lake Jordan which in the Ice Age filled the rift valley. The flood plain, known as the Zor, is about four miles wide except where it opens out into the broader valleys of Beth Shaan and Jericho. In the course of the 65 miles from the Sea of Galilee to the Dead Sea the Jordan River falls from 680 feet below sea level to 1290 feet below sea level. In this course the river follows a remarkable series of meanders that increases the length of its flow to 200 miles.

The Dead Sea

The Dead Sea, lying at the deepest part of the great Jordan inland depression, is 47 miles long and 10 miles wide with an area of about 360 square miles or 230,000 acres. The mean depth is 1080 feet and the greatest depth 1310 feet.

Evaporation from the Dead Sea under furnace-like temperatures is extraordinarily high. It has been estimated at from about six and a half[2] to seventeen feet[3] a year, and definite studies must be made to determine the average evaporation over this 230,000-acre surface of water. This great quantity of evaporated moisture is believed to account for the peculiar blue-white clouds that form ethereal mists over the surface of the Dead Sea, as we noted in our two

[2] M. J. Ionides, *Report on the Water Resources of Trans-Jordan*, p. 143, 1939.
[3] Ch. Audebeau Bey, *La Vallée du Jourdain*. Ministère de l'Agriculture, Direction des eaux et du génie rural, *Rapports et notes techniques*, fasc. 57, 1927, p. 413.

visits to that eerie region. This great loss of water is restored mainly by the inflow of the Jordan water and by the discharges of some small wadies, mainly from the East.

The climate of this region prior to the Ice Age was similar to that now prevailing, and the Dead Sea was of approximately the same size. During an era of heavy streamflow, called the Pluvial period, the surface of the Dead Sea was raised by inflow greater than evaporation loss until the Pleistocene lake, appropriately called Lake Jordan, reached a height 1400 feet above the present level and higher than the Mediterranean. But the natural dam of the Araba, 1200 feet above sea level, prevented the waters of Lake Jordan from escaping. At its largest extension, Lake Jordan was two hundred miles long, from Huleh in the north to a point forty miles beyond its present southern limit. That life could exist in these waters is shown by fossils in their deposits.

When runoff of streams was limited to that of annual precipitation, the evaporation loss was greater than the inflow and the waters of Lake Jordan gradually shrank to the present Dead Sea, just as the waters of Lake Bonneville of Pleistocene times shrank to the Great Salt Lake of Utah today.

The level of the Dead Sea even in our times has not remained constant. From 1879 to 1929 it rose 23 feet, and from 1929 to 1935 it fell 8.75 feet.[4] It is presumed that on the whole the Dead Sea has been rising for centuries. This, however, does not necessarily mean that rainfall has increased,

[4] M. A. Novomeysky, *Trans. Institution Chem. Engineers*, Vol. 14, 1936, p. 60.

but may only mean that the coefficient of runoff has increased because of less absorptive condition of the drainage.

The salinity of the Dead Sea water is four to five times that of ocean water, namely, 23 to 25 per cent of salts in Dead Sea water against 3 to 6 per cent in ocean water. Analyses made of the Dead Sea waters at various levels disclose that the salt concentration increases with depth. The accompanying table gives an analysis by Terreil taken at a depth of 390 feet below the surface, where the percentage of solids is 24.5. The figures in the table show percentages in the total weight of solids.

ANALYSIS OF DEAD SEA WATER 390 FEET BELOW SURFACE

Chlorine	67.66	per cent
Bromine	1.98	" "
Sulphate	0.22	" "
Sodium	10.20	" "
Potassium	1.60	" "
Calcium	1.51	" "
Magnesium	16.80	" "

Magnesium, potassium and bromine, minerals which are now in great demand, are contained here in vast abundance. These salts have been accumulated by evaporation of the inflow of salt-charged waters. Bitumen floats ashore on occasion. Josephus called the Dead Sea the Asphalt Sea, suggesting that in his time recovery of asphalt was an important occupation.

No other sea has been called by so many names or had such a hold on men's imagination. Wild tales told by medieval travelers say that no plant can live in the poisonous air of the Dead Sea, that no birds fly through it and no wave disturbs its glossy surface. Actually, even people of European extraction can live on the shores of the Dead Sea for years without harmful effects. Its winter climate is one of the finest in the world. The scenery along the south shore is beautiful and romantic and makes the Dead Sea area an ideal site for winter resorts. The intense coloring of the sea, the varied effects of light, the overhanging slopes broken by deep gorges produce a wild and sublime beauty.

The remarkable features of the Jordan depression make possible a great irrigation and power plan which, when developed, may transform Palestine into the industrial center of the Near East. A fuller description of this plan is given in Chapter XI of this book, dealing with the "Jordan Valley Authority."

Climate and Rainfall

The Jordan Valley and the maritime slopes of Palestine present a great variety of climate, soil, flora and fauna. Palestine's first High Commissioner, Lord Herbert Samuel, remarked justly that the "country offers, on an area of a province, the soil and conditions of a continent." The climate is cool in the northern hills and definitely tropical in the deep depression of the Jordan. While certain districts of Greater Palestine are suited to all crops and fruits character-istic of the temperate zone, most of the other areas are too warm for successful cultivation of deciduous trees, but are excellent for olives, figs, grapes and citrus fruits; still other sections are suitable for palm trees, bananas, avocadoes and other tropical plants. When one leaves the scorched shores of the Lake of Galilee on a hot summer day and after an automobile trip of only one hour reaches the town of Safed in the hills of Galilee, one feels as if he had entered a totally different country.

[35]

The range and distribution of air temperature place Palestine as a whole in the sub-tropical belt, though areas like the Ghor and the Dead Sea experience tropical heat, and high mountain elevations enjoy the coolness of a temperate climate. Freezing is not uncommon in Jerusalem (2900 feet above sea level) in winter, but snow is infrequent and rarely lasts more than a day. To the north, heavier snowfalls occur, and on Mount Hermon perpetual snow lies on the protected flanks of the peak.

In ancient Hebrew lore, the year was divided into only two seasons—the autumn and spring season of the rain, and the season of the sun, for there are no distinct transition periods. Rainfall is generous in the highlands during the winter but is slight in the Negeb in the lower part of the Jordan Valley and in the Sinai Peninsula where it drops to five inches and less per year. Rainfall in these regions is irregular so that droughts may occur today as they did in ancient times.

The season of the sun is one of great summer heat tempered by fresh westerly breezes off the Mediterranean Sea during the day, and by cooling winds out of the desert in the night. The nearness of the great Arabian desert to a warm subtropical sea, and their separation by the Palestine and Trans-Jordan highlands, brings about a remarkable daily alternation of winds. During the day as the desert heats up under the blistering rays of the summer sun, the air rises and draws in winds from the sea. During the clear, starlit summer nights the desert cools off more quickly than the sea, and the air rises over the Mediterranean and pulls winds off the desert across the mountainous lands of Palestine and Lebanon. In narrow valleys and mountain gaps, winds rage and blow at high velocity as if trying to free themselves from a prison, and raise dust clouds into the sky.

It was owing to strong winds of this kind that Solomon was enabled to build copper smelters at Ezion Geber on the Gulf of Aqaba 2800 years ago. Dr. Nelson Glueck has recently brought this ancient "Pittsburgh of the Red Sea" to light. At the time of our visit to his excavations a strong wind blew dust into our eyes, and gave us an idea of how workers at these ancient furnaces depended on such a wind to fan the flames of their smelters.

An exception to this daily seesaw of winds from sea to desert occasionally occurs in the spring when the east wind blows in the daytime. This is the dreaded "khamsin" which is like a hot blast of a furnace filled with fine impalpable dust that irritates the mucous membranes of animals and humans. East winds wilt vegetation unless it has depth of soil.

While at Kurnub in the Negeb, inspecting the ruins of ancient dams and irrigated terraces, we watched a "khamsin" moving westward like a great wall and overcoming the west wind off the sea. As the "khamsin" bore down upon us, the wind changed from west to east, and thereafter the atmosphere was full of irritating dust.

During the season of the sun, as a rule, there is no rain from May to November. The season is long and dry, except for the tempering effect of dew which reduces evaporation losses and adds to the moisture supply for vegetation. Understanding this occurrence of dew, Gideon of Israel called for dew to be deposited on a fleece when the ground was dry, and again for dew to appear on the ground when the fleece was dry (Judges 6:37-40). In recent years scientific studies by Lebedeff of Russia have shown how the conditions that produce dew also bring about condensation of moisture within the soil.

In October, after a long, dry season the land is thirsty and ready to drink in the blessings of heaven. Moreover, the land is naked, for flocks of goats and sheep have eaten all palatable vegetation into the ground. We saw the hungry goats in autumn attempt to climb trees to get at the leaves for forage.

The season of the rain is the time to store water—to store it in the soil and in cisterns and underground. The farmer plows the land to open it up to the rain. Gentle rains are welcomed because they are fully stored in the soil; when rain comes in heavy dashing downpours on the parched earth, much of it runs off quickly with a corresponding loss of benefit. The character of the rain is thus of primary importance to the success of crops. By the use of measures for conserving soil and water, it is possible to store hard rains as well as gentle ones.

The most beneficial winter rains are those which are evenly spaced. If rains occur at long intervals, crops suffer unless they are planted in soil deep enough to store moisture. It is for this reason that the "latter rains" are important. They replenish the moisture of the soil sufficiently to carry the crops through to harvest time. We experienced such a "latter rain" on February 21 and 22, 1939. It came straight down in a steady and heavy fall, yielding an abundance of water. Most of the hill land, however, was not prepared for this generous blessing. Storm runoff was very heavy, and all drainage channels were filled with ugly currents of dark-brown gully washes, indicating that soil was being eroded and carried away from the slopes where the soils are already too thin. Gone with the rain was much soil from the slopes.

The annual amount of rainfall not only varies greatly in different districts, reaching its maximum of seventy inches on the slopes of Lebanon and its minimum of five inches in the Jordan Valley and Sinai Peninsula, but it also shows wide variations from year to year in the same locality. These variations are greatest in the semi-arid Negeb and impose peculiar problems on farming in that region. We have, however, encountered the same problems in the great southern plains of the United States, where we have worked out a solution that may be adapted to the Negeb.

Similarities to California

Generally speaking, the climate of Palestine resembles that of Southern California to a remarkable degree. Both have sharply divided rainy and dry seasons. The average annual temperature of the Palestinian coast is very near to that of Los Angeles, although in Palestine the difference between the temperatures of the summer and winter is more pronounced. The average yearly temperature of Tel Aviv is 67° F. compared with 62.5° F. of Los Angeles.

The total annual rainfall in Los Angeles is less than in Tel Aviv, but the distribution through the year is strikingly similar. In both cases over 90 per cent of the total rainfall occurs in the six months from November to April. The four months from June to September are actually rainless, while the precipitation in May and October is negligible.

Although Southern California is much larger in area and structure than Palestine, it is a remarkable counterpart in many respects of the Holy Land. The Colorado River with its muddy water is much like the Jordan. The Salton Sea, below sea level and rich in salts, is much like the Dead Sea. The Mountains of San Gabriel topped with Mount Baldy remind one of Mount Hermon, and the San Bernardino Mountains remind us of the Jebel Druz of Trans-Jordan. The valley floor of the Santa Ana River is like the Maritime Plain of Palestine, in which irrigation is from underground waters replenished by storm flow of small streams issuing out of mountains. Moreover, the Jordan Valley offers opportunities for developing as unique and remarkable a power project as Boulder Dam and the Los Angeles Aqueduct, but at a much lower cost.

The outstanding difference between the two areas is their geological structure, and in this respect Palestine is more favored. The formations of limestone underlying its hills dissolve into labyrinths of underground channels, which take in much of the winter rains and lead them in time to come forth as great springs in the depths of the valleys. On the other hand, Southern California, with its complex of igneous and gneissic rocks which are impenetrable to rain water, does not possess big springs in its mountain valleys, and the residual soils on the slopes are too thin for cultivation.

How much water is discharged in submarine springs in Palestine is not known, but the quantity may well be considerable. Most of the runoff is through wadies, which, dry in summer, become floods during heavy rainstorms. During the continuous rains of 1926-27, Wadi Surar in the Maritime Plain discharged 300 cubic feet per second, and Wadi Kelt near Jericho was a raging torrent of 300 to 450 cubic feet per second. Any systematic program for developing water

[40]

resources in Palestine will have to take full account of these facts.

With the exception of the Jordan, the Yarmuk and the Yarkon (Auja) rivers, the water courses of Palestine actually disappear in the summer, although they carry considerable amounts of water in the rainy months. Most of them have their source in the hills of Samaria and Judaea, and carry their water to the Mediterranean Sea, whereas those of the Trans-Jordan plateau discharge into the Jordan River and the Dead Sea and are eventually lost by evaporation.

The open streams of Palestine issue forth from springs whose discharge is regulated by the internal underground reservoir within the labyrinths of limestone rock. Some flow in the rainy season only, indicating a more local and restricted underground reservoir. Others are all-year springs, showing little influence of the rainy season; such are the great springs of the Auja, with a discharge of 291 cubic feet per second, and the springs of Beisan discharging 51 cubic feet per second.

The existence of springs greatly facilitated human occupation of the mountain fold along the eastern Mediterranean. Since wells in the limestone are rare, springs were necessary to permanent settlement in ancient times, until, according to Glueck, the cistern was invented in the Iron Age as a way to store water for the long rainless season.[5] The water of many springs is used for domestic supplies and for irrigation of the more valuable crops of vegetables and orchards; however, grain crops, olive and fig orchards, and vineyards are usually watered by winter rains alone.

[5] Nelson Glueck, *The Other Side of the Jordan*, p. 121. New Haven, 1940.

Soils of Palestine and Trans-Jordan

In the formation of the soils of Greater Palestine rock, climate, former vegetation and topographic features influencing erosion were the outstanding factors. While the soils frequently show a great variety even in neighboring fields, the country can be generally divided into seven topographic districts. They are, according to Reifenberg:[6]

(1) The Maritime Plain
(2) The Highlands of Judaea and Samaria
(3) The Rift Valley of the Jordan, Dead Sea and Araba
(4) The Emek, including the Plains of Esdraelon and Beisan
(5) The Highlands of Galilee
(6) The Negeb, a plain merging into the desert of Sinai
(7) The Highlands of Trans-Jordan

(1) *The Maritime Plain*: The Maritime Plain was formed by fossil sand dunes, old alluvium and recent alluvium brought down by torrents from the eroding slopes of the highlands and spread out over a base complex of Tertiary limestone and calcareous sandstone marls. Fossil sand dunes are solidified by calcareous cement. The Maritime Plain received surface and underground runoff from the highlands. A gray sandy-clay horizon at a comparatively shallow depth keeps the ground waters within well depths and makes possible irrigation pumping at low costs.

Following the breakdown of terrace agriculture in Palestine beginning with the seventh century A.D., erosion of soils from the slopes of the highlands brought down an enormous amount of material that choked stream channels, dammed up drainages which had flowed freely in Roman times and caused marshes to spread over low-lying flat areas. Malaria became pestilential until Jewish settlements in recent times

[6] Adolf Reifenberg, *The Soils of Palestine*, London, 1938.

[42]

started to drain the swamps. When drained and irrigated, such accumulated soil material turns into some of the most productive districts of Palestine where intensive agriculture can be developed to produce high yields.

North of the Carmel spur, the soils of the Maritime Plain are alluvial loams. South of Carmel in the Plain of Sharon are found red sandy soils in which citrus culture thrives. These soils extend to an east-west line through Rehovoth, south of which is a group of Mediterranean steppe soils as far as Gaza. All along the coast the Maritime Plain is bordered by sand dunes in varying degrees of fixation. Northward of Gaza sand dunes are active and advancing on farms, orchards and villages.

(2) *Highlands of Judaea and Samaria:* Country rock of the highlands bordering on the Maritime Plain to the east is limestone lightly folded and heavily faulted in a block anticline. The soft chalky Senonian limestones between Nablus and the Valley of Esdraelon give rise to silty loams, porous and grayish in color with low natural fertility. They suffer little from erosion as compared to soils derived from other formations.

Soils derived from the hard Turonian and Cenomanian formations that make up the highlands of Judaea are remarkable red earths whose origin is still a subject of lively discussion. Called Judaean clays by Strahorn, they are heavy clays, rich in lime, colored deep red and maroon by ferric iron. Highly aggregated into a crumb structure, they are suited to cultivation.[7]

Erosion takes a fearful toll from the uplands during heavy

[7] A. T. Strahorn, Joint Palestine Survey Commission, "Soil Reconnaissance of Palestine," *Reports of the Experts*, pp. 143-236, 1928.

winter rains, but fortunately the limestone country rock is shot through with solution fissures and pockets that are filled with red earth. This feeds trees and shrubs and makes forest tree plantations still successful, in spite of the enormous erosion that has taken place there since the ruin of the ancient terrace culture.

The valleys on the western slope of the north-south highland range of Palestine are narrow, steep-walled gorges that cut through the steep-faulted wall of the highlands. Soils on these slopes are lacking or are thin and rocky. In winter the gorges are filled with raging torrents during heavy rains, but in summer they are dry.

The east wall of the highlands is formed by faults of the great rift valley and falls sharply from elevations of 2000-2900 feet above sea level to the floor of the Jordan depression, a total of 3200 feet in a distance of 15 miles from Jerusalem to Jericho. The drainage channels are deeply cut in steep-walled valleys, showing that active cutting is continuing. Down the steep escarpment slope the red soils of the highland plateau fade out rapidly in color to a dull gray that tints the landscape of the Jordan Valley.

The plateau of the highlands was once covered by a mantle of red earths of ancient origin. Evidence is found in outwash deposits in the Jordan Valley where bands of this red earth are spread out. This plateau was once covered with groves and forests. Soils here have been intensely cultivated to rain-watered crops, first mainly to grain, and then to orchards, and have suffered enormous losses by erosion.

Little or no irrigation is possible in the highlands. Springs are rare, for internal drainage lets water drop to levels far below the surface. Wells deep enough to reach ground

[44]

water would be justified only for domestic and municipal use. That this condition has prevailed for a long, long time is shown by the general use of cisterns which dates back to the Iron Age.

The soil profile is full of weathering hard limestone rocks that supply excellent material for building terraces. In Byzantine times, before the Arab conquest, terracing in the highlands was widespread and well done. Crops planted on these irrigated terraces were chiefly tree crops, such as olives, figs, pomegranates, vines and carob trees. Terraces, however, are effective only when their walls are carefully maintained. When they were allowed to fall into ruin, the terrace agriculture broke down.

In spite of their desolation, the highlands of Palestine can still play an important part in the rebuilding of the country. Even where remaining soil, after centuries of erosion, is too thin for grain or vegetable crops, this extensive area is suited for grazing and forestry management as well as for certain tree crops which may improve the food balance of the country directly or indirectly.

(3) *Soils of the Jordan Valley*: The Jordan Valley, as the lowest inland depression on earth, has an unusual interest for geologists. The "Lisan" marls are Jordan Lake deposits exposed by recession of a Pleistocene lake, and these beds have been cut into by cross-flowing wadies.

In the northern part of the Jordan Valley, in the Huleh area, the soils are alluvium washed in behind the basaltic lava dam, and are in need of drainage for their highest use. Soils of this area are heavy and best suited to grains, forage crops and vegetables.

To the south of the Lake of Galilee the Jordan Valley

[45]

consists of a broad plain to about ten miles below Beisan, in which the Jordan River has developed a flood plain of varying width that lies thirty to sixty feet above the present floor. The river meanders from side to side in this flood plain, cutting into the steep banks and carrying eroded material downstream into the Dead Sea. Active cutting of the river's course in this area makes the Jordan water muddy at all times of the year, even through the dry season when no erosion takes place.

The possibility of irrigating Upper Jordan lands with water from the river itself and from the Beisan springs has been demonstrated by the Dagania colony. Extension of irrigation in this area is possible on a still larger scale under a comprehensive program to use all the fresh water of the upper Jordan. The soils here are dark-brown silt loams, with good drainage and free of alkali.

The lower Jordan Valley, extending from Tel Abn Sidreh south of Beisan to the Dead Sea, a distance of little over thirty miles, is naturally divided into a narrow flood plain along the river, the river terrace of sediments left by Lake Jordan of the Ice Age and alluvial fans that have been spread out over this terrace from wadies on both sides of the valley. The flood plain called the Zor is inundated by annual floods and, according to Strahorn, its soils are alkaline. The sediments once laid down in Lake Jordan of the Ice Age are marls extending from the base of the Judaean hills and the Trans-Jordan escarpment, into which the Jordan River has cut its flood plain.

The land is now suitable for agriculture only where sediments eroded by wadi streams from the highlands on both sides of the river are spread out over the marls. Experiments, however, indicate that the alkaline soils from marls may become suitable for intensive farming after they are leached of salts with fresh stream waters.

Because of the low rainfall and great heat of the Jordan Valley, irrigation is required for cultivating these areas. Outwash fans of the wadies, being the best soils, are also subject to occasional if not yearly floods quite similar in nature and intensity to those of Southern California, where highly productive citrus orchards and vineyards are planted on outwash fans along the Piedmont in the San Gabriel and San Bernardino Mountains. As in California, measures of flash flood control are necessary to protect these lands when cultivated. The reclamation and continued use of these good soil areas will call for regulation of the entire drainage systems from which storm flow comes. Irrigation waters must be free enough of alkali to prevent accumulation of salts in soils now suitable for crops.

The extent to which the Jordan Valley may be reclaimed, if measures are carried out to leach the soils of salts, cannot be fully known until an intensive survey is made. It is, however, estimated that about one-half of its total area of approximately 500,000 acres can be reclaimed by leaching, by terracing and by irrigation and drainage, thus making available a total of 250,000 acres for intensively farmed land. This does not include slopes of the west and east walls of the Jordan Valley, which, when irrigated by a high canal as designated by the "Jordan Valley Authority" reclamation plan, may be converted into terraced orchards and vineyards or utilized for grazing livestock.

(4) *The Plain of Esdraelon*: From Acre and Haifa southeastward extends low-lying land that widens into the historic

Plain of Esdraelon, which, broken in its midst by a low ridge, continues farther to the east where it merges into the valley of Beisan outbranching from the Jordan Valley. It was known in history as the main road for travel between Egypt and Damascus, and beyond to Mesopotamia. In recent times engineers built the railroad to Damascus along the same route.

In ancient days the rich lands of Esdraelon were a source of forage for animals and of food for caravans and armies. The soils of these low-lying areas are generally heavy, and, until drained by Jewish colonization, they were marshy and malarial. All soils of this area are of ancient origin except around the margins, where recently eroded material has been brought down out of the highlands and spread out by winter floods. The soils are generally subject to deep and wide cracking in the dry season. On the whole, the old alluvial soils of these valleys are suited to grain crops, without irrigation, but they require irrigation for full utilization for vegetables, tree crops and vineyards. With irrigation waters from the Upper Jordan, the valleys of Esdraelon and Beisan may be a veritable garden land where farmers can make a living from small intensively cultivated plots.

(5) *The Highlands of Galilee*: North of the Emek the land rises into the highlands of Galilee, in a series of hills that gradually ascend to the dominating range of the Lebanon Mountains. The country rock of the hills of Galilee consists chiefly of hard limestone that weathers into red earths, and of volcanic basalts that broke through and flowed over the limestone mantle rock. The basalts give rise to a series of fertile soils whose cultivation, because of steep slopes, would require the restoration of ancient terraces.

[48]

Enjoying the climate of the Levantine coast, the highlands of Galilee receive a heavier rainfall than most other districts of Palestine. While heavy rain has taken its toll of soil by erosion where old terraces have fallen into ruin, springs and ground water furnish a year-long water supply for domestic and irrigation purposes.

The valley floors of Galilee are filled with offwash to depths that make agriculture profitable. Reclamation of hill slopes which is being begun in Hanita, Dafne and a few other recently founded Jewish settlements calls for building terraces, at great cost in labor, and planting orchards on them. The settlers of Hanita, while clearing ground, dug into the roots of some very large olive trees, indicating the former use of these lands. Galilee has long been noted for its olive groves, of which few now remain.

Soils from weathered igneous rocks are fertile clays, but steep slopes make the cultivation of these lands under the heavy and torrential rains of northern Palestine hazardous because of erosion. However, they may be utilized and rationally developed for grazing and reforestation.

(6) *The Negeb*: The southern continuation of the Judaean hills and the hilly plain of the Shephelah is a semi-arid plateau sloping generally to the south and west, known as the Negeb (Hebrew for South). Though it contains near half of the total area of Palestine, it is the most desolate and sparsely populated part of the country. However, the ruins of ancient cities and remnants of irrigation works which have been found there arouse hopes for the possible restoration of this much discussed area's former prosperity.

The country rock of the Negeb is generally mantled with loess and sand dunes. On the road from Beersheba to Kurnub

we examined a section showing chalky limestone overlain
by twenty-six feet of coarse loess. During a plane flight
over the Negeb, I saw distinctly the blanket of loess spread
out over the central part of this area. To the west lay a zone
of brownish sand dunes. Later a ground study convinced
us that the dunes are now active, rolling across the plain
and menacing fertile lands. But to the east the loess blanket
gives out and exposes an old landscape surface in which
head cutting by wadies (seasonal streams) is taking place.
The soils of this great area are fertile when sufficiently
watered by rain or by irrigation ditches. The prospect for
restoring the Negeb to its former state of prosperity will
be examined at length in Chapter XII on the Possibilities
of the Negeb.

(7) *The Highlands of Trans-Jordan*: No soil survey has
been made of Trans-Jordan. According to our own observations the soils in the highlands are red earths similar to
those of the Judaean highlands; they have suffered heavily
from soil erosion where they were once cultivated. Out
toward the desert in the east and to the south, soils turn
yellowish and grayish and, under lessened rainfall, can be
used only for grazing.

So important, however, is the area of Trans-Jordan that
it is treated at length in the further course of this book
(Chapter XIII, "Trans-Jordan—Past, Present and Future").

Analyses of Palestine soils made at the Experimental
Station at Rehovoth have shown that with the exception of
the Beerin series in the Rehovoth area, and the loess soils
of Beersheba, sufficient phosphorus is present. However,
there is a deficiency in nitrogen in all soils with the exception
of those of the Upper Jordan series and those in the Plain

of Esdraelon. Likewise, owing to the inadequate and wasteful methods of native farming that have prevailed for so many centuries, the soils of Palestine are deficient in organic matter. However, the soils of the Jewish colonies have quickly responded to fertilizers and indicate a general improvement.

Natural and Cultivated Plants

Much has been said and written about the original vegetation of Palestine. The prevailing assumption is that Palestine was once covered with forests. But this cannot apply to the whole country. The amount of precipitation varies widely from place to place, ranging from desert to humid conditions, and it is natural that vegetation should vary in the same way.

Because of the remarkable topographical and climatic similarities between Palestine and California, we may reconstruct the vegetation of Palestine by comparing it with that of California, where vegetation after 150 years of occupation is much less disturbed than it is in Palestine after 1500 years of intensive cultivation and 1300 years of neglect.

In its natural vegetation, the Holy Land is similar in most respects to California. The paloverde tree of the Mojave and Coachella Valley deserts correspond to the shittim trees (*Acacia Seval*) of the Sinai Desert. The thornbushes of Palestine, such as camel's-thorn (*Pittosporum sponosum*) and Christ's-thorn (*Zizzyphus Spina Christi*) correspond to the thorny, semi-desert vegetation of mesquite and catclaw in California. The grassland savannah of the highland areas of Southern California corresponds to the seasonal grass and herbage cover in the region about Beersheba and in the highlands of Trans-Jordan. The evergreen oak woodlands still to be found along the foothills of Southern California resemble the savannah of the Maritime Plain of Palestine, where mystic forests of oaks once attracted the attention of the Crusaders and Napoleon's army. The chaparral or elfin forests of Southern California and Arizona remind us of the leather-leaved evergreen brushy growth of terebinth and of pistachio shrubs and dwarf oaks of intermediate altitudes in Palestine. And finally, the forests of pine and big-cone spruce and the big redwoods of California correspond to mountain forests of pine, cypress and cedar of Lebanon.

Both in California and in Palestine we find similar types of vegetation in similar climatic zones. We should not, therefore, expect to find forests in the desert areas of Palestine which are similar to the deserts of California. On the other hand, where we find forests in California, we should expect that forests grew in similar climatic zones of Palestine. And in between the forests and the grasslands zone, we should expect to find the woodland type of vegetation—groves of trees with open grassy parks between.

In Palestine and Trans-Jordan we did indeed find remnants of ancient forests which once covered considerable parts of the hilly country. In the Maritime Plain we studied remnants of the oak forests referred to by the Crusaders and later by an officer of Napoleon's invading army. In Trans-Jordan we examined stumps of trees in a forest that grew at an elevation of about 5000 feet between Kerak and Ma'an. This forest was cut clean during the First World War by the Turkish army; lacking coal, it used the wood as fuel for railway locomotives. In Trans-Jordan we likewise examined the forest of Ajlun which still covers several thousand acres of land. And in Lebanon we studied the remnants of the famous cedar forests.

One of the outstanding species in the original forests of Palestine was the Aleppo pine (*Pinus Halepensis*), which grew on the heights of Mount Carmel and on the highlands of Judaea and Galilee. The plantation of this pine on Mount Scopus near Jerusalem, in the Balfour Forest and other reforestation areas developed by the Jewish National Fund indicates that it will thrive again in the highlands of Judaea, Samaria and Galilee. Species of oak, like the Tabor oak (*Quercus Ithaburensis*) made up the woodlands in the Maritime Plain; remnants of them may be found along the Haifa-Nazareth road to the north of the Jewish village of Nahalal.

Palestine possesses about 1500 species of wild plants. Many beautiful flowering plants bloom in the winter and spring to make the land a place of beauty even as California is in the springtime. Dr. Ha-Reubeni of the Hebrew University has made a complete and remarkable collection of Palestine's flowering plants. The anemone is the most widespread flower on the fields of Palestine; it covers valleys and hill slopes with a carpet of red. *Cyclamen persicum, Ranunculus asiaticum, Tulipa montana* and many gorgeous kinds of wild iris beautify the countryside in the spring. After June, most of Palestine's annual plants are dried up, and by the end of summer the country, with the exception of cultivated and irrigated fields, presents a desolate appearance. But soon after the November rains, the entire landscape again turns green, as it does in California.

The cultivated plants of Palestine and Trans-Jordan have unusual interest and value. Aaron Aaronson, the great botanist, found samples of wild wheat in the Hauran and on Mount Hermon which led him to believe that wheat had originally come from this part of the Near East. Wheat and barley were developed by the early farmers of the Near East into valuable food plants that became the staff of life of ancient civilizations, and of the Western World. Other plants such as lentils, sesame, kersanna and certain vegetables—leek, onions, cucumbers, melons, etc.—have long been grown in the Near East.

The tree crops have a special place in these lands. The fig, the olive and the vine were traditionally most important sources of food. The olive tree grows wild and it is probable that cultivated varieties were developed from wild strains of olive trees on the mountains. The fig has played an important part as a food plant, and so has the sycamore fig tree which furnished a coarse food and grew from Egypt to the Jordan Valley. The vine has long been a crop grown in the highlands of Palestine and figures in Palestine's literature from the earliest times.

Of other tree crops, the carob tree (*Ceratonia siliqua*) deserves special mention. Its pods, rich in honey, were the husks in the famous story of "The Prodigal Son." They have a high food value as a browse and forage plant for livestock. Among the nut trees of Palestine are the almond, walnut and pistachio. Important foods are supplied too by the apricot and pomegranate trees and by the date palm which once grew throughout the Jordan Valley.

Citrus trees are not native to Palestine and are not mentioned anywhere in the Bible. The orange tree, of Far Eastern origin, was introduced in the Middle Ages by Portuguese traders; its Arab name, *Burdugan,* is a corruption of "Portugal." It is now the most important tree crop of Palestine. A wide variety of other plants, as well, are suited to the climate of Palestine and can produce food and forage for the land.

Mineral Resources

The most important mineral resources of Palestine are those contained in the highly saturated water of the Dead Sea. The quantities of common salt (sodium chloride), potassium chloride, chloride of magnesia and magnesium bromide dissolved in the water of this remarkable lake are practically inexhaustible. According to experts, the Dead Sea contains 2000 million tons of potassium chloride, 980 million tons of magnesium bromide, 11,000 million tons of sodium chloride, 22 million tons of magnesium chloride and 6000 million tons of calcium chloride. Large quantities of potash and bromine are being extracted by the Palestine Potash Syndicate, a company owning the concession for the exploitation of those important minerals, and the extraction of magnesium, extremely important in the conduct of the present war, is about to be started. Besides, the mineral resources of the Dead Sea are increasingly used as a source of fertilizers and as a foundation for the creation of a chemical industry. Superphosphates, chlorine, sulphuric acid and caustic soda are some of the products manufactured in connection with the reclamation activities on the Dead Sea.

Other important Palestinian mineral deposits are sulphur in the southern part of the country and phosphate rocks in Trans-Jordan, near the frontier of Mandated Palestine. Asphalt and large deposits of bituminous limestone are further possibilities. Rock salt is mined at the Jebel Usdoom, where, according to a local Arab legend, Lot's wife was transformed into a mountain of salt.

The rocks in the vicinity of the Dead Sea contain considerable quantities of oil, but whether the petroleum deposits of Palestine could be successfully exploited for commercial purposes in competition with the rich oil fields of Iran and Iraq is yet to be decided.

Palestine possesses a variety of valuable building materials such as marble, basalt and gypsum. It has sizable deposits of shale and lime, serving as the foundation for a comparatively large cement industry. Some of its earth deposits are naturally suited to the production of aluminum, but the development of such an industry depends on cheap water power, which is in sight in the Jordan Valley Authority power project.

Palestine possesses a number of mineral springs whose healing properties were highly esteemed in ancient times. At the time of the Romans, three bathing spas—Tiberias, Gadara and Kallirhoe—were famous beyond the frontiers of the country. Besides, the water of the Dead Sea seems to possess definite curative qualities. Extracts of their minerals are now being used by the British armies of the East in treating wounds.

CHAPTER IV

PALESTINE IN ANCIENT TIMES

. . . .

*"A land wherein thou shalt eat bread without
scarceness, thou shalt not lack anything in it."*
—DEUT. 8:7-9

HISTORICAL data and archaeological finds furnish abundant
evidence of a remarkable period of prosperity in Palestine
which lasted for more than a thousand years before the
decline that began in the seventh century A.D. It will be
helpful in our study of the possibilities of Greater Palestine
to know something of the country's density of settlement and
economic life in the past.

The population of Palestine and Trans-Jordan in ancient
times was far in excess of their present population of
1,800,000. Historians differ in their estimates, but they all
agree that from the sixth century B.C. up to the destruction
of the Second Temple in 70 A.D., Palestine had a much
larger population than at present. The population seems to
have reached its peak during the first century of our era.

Juster Armunk, one of the most responsible historians of
Palestine, estimates its population in the time of Christ at
five million (including Trans-Jordan). Josephus Flavius
suggests a still higher figure, but modern historians believe
that Josephus tended to exaggeration. Avi-Yonah, a modern

PALESTINE, LAND OF PROMISE

historian living in Palestine, believes that five million inhabitants should be taken as a minimum figure.[1] Dr. Maisler,
another modern historian, states that latest archaeological
investigations indicate that the population of Palestine was
in excess of four million.

More conservative estimates are given by other students.
Professor Salo Baron of Columbia University estimates
Palestine's population at the time of Christ as between
2,500,000 and 3,000,000 but he admits that his estimate is
conservative. Dr. Nelson Glueck, guided by his own archaeological discoveries, estimates the population of Palestine
during that era at about 2,500,000, which he also considers
a conservative figure.[2] F. C. Grant reaches the conclusion
that there were approximately 250 inhabitants to the square
mile in western Palestine and somewhat fewer in Trans-
Jordan.[3] This would indicate a population of 2,500,000 for
the present Mandate area, and at least one million for
Trans-Jordan. The fact that Jewish casualties in the Roman
wars are said to have run into the hundreds of thousands
would indicate a fair-sized population to begin with. Tacitus,
a cautious historian, sets the figure for Jewish casualties at
600,000. Added to this evidence is Adolph Buechler's con-
tention that Judaea remained fairly well populated even
after the Roman wars, despite the heavy losses sustained in
their course.[4]

[1] "Middle East's Economic Past," *Palestine and Middle East Magazine*,
Vol. 11, 1939, p. 87.
[2] Letter to the American Economic Committee for Palestine, dated December 16, 1941.
[3] F. C. Grant, *The Economic Background of the Gospels*, New York, 1926.
[4] Adolph Buechler, *Economic Conditions of Judaea After Destruction of
the Second Temple*. London, 1912.

For the purpose of this study, it is best to err on the side of conservatism. We may conclude, therefore, that a figure of three million inhabitants for Palestine and Trans-Jordan at the beginning of the Christian era would be acceptable to the most conservative sociologists and archaeologists.

Two thousand years ago Palestine, with its neighbors, Egypt and Syria, represented one of the most advanced economic areas of the Roman Empire. By virtue of its favorable geographic position, the Near East became a flourishing industrial and agricultural center. Palestine, situated on the crossroads of trade routes from India to Rome and from Central Asia to Egypt, served as a land bridge between the three continents of Europe, Asia and Africa. An inland route extended from Syria down the Jordan Valley and Arabia to Egypt, and another land route followed the coast. An east-west trade route extended from the northern arm of the Red Sea at Aqaba through southern Palestine to Gaza on the Mediterranean coast. The latter became the main outlet for trade passing from Arabia to points west, principally Greece and Rome. Palestine has not been blessed with natural harbors, but Acre and Joppa (now Jaffa) served her commerce in ancient times. To supplement them Herod built a port at Caesarea and improved Gaza.

Although slave labor made mass production possible in antiquity, mass transportation was limited. Ancient ships were small and dependent on sails or galley slaves for power. Overland transport by horse, donkey or camel did not permit the carrying of bulky goods. Thus merchants had to concentrate on transporting articles of the highest possible value and the least bulk and weight. Long-distance trade, which extended from the Far East and India by way of the Near East to Europe, both by land and sea, was therefore limited largely to luxury products, such as silks, spices, carpets and perfumes. Palestine's position on both land and sea trade routes brought much wealth to the country. Usually, Arabian caravans and Red Sea shippers sold their goods to Syrian and Palestinian merchants, who, in turn, forwarded them to Greece and Italy.[5] Pliny estimates the annual balance of eastern trade adverse to Rome at nearly five million dollars, a large sum at that time. A goodly part of this must have represented the profits of middlemen, and the Palestine merchants undoubtedly had their full share.

The Decapolis, or the League of Ten Greek Cities, established in Trans-Jordan about 300 B.C., provided security and increased opportunities for trade. After the conquest of Palestine, Rome built splendid roads, some of which still exist. On our way to Petra, the ancient Nabatean capital, and Aqaba, Solomon's seaport on the northern arm of the Red Sea, we crossed and recrossed such highways, their well-laid paving stones worn down by many chariot wheels.

Roman Palestine formed part of an extensive territory possessing a unified currency. Only Roman gold and silver were allowed to circulate. The whole of the Roman Empire was a single free-trade region divided for fiscal purposes into customs blocks. At frontiers only a moderate charge was levied on goods passing through. The absence of tariff protection in ancient times was favorable to the eastern half of the Empire, with its highly developed industrial and commercial centers. There was no organized system of commercial banking in antiquity, but the royal treasury under

[5] M. Avi-Yonah, "Economics of Ancient Palestine," *Palestine and Middle East Magazine*, Vol. 9, 1937, p. 189.

Herod financed trade ventures in Palestine. Also, the Temple at Jerusalem accepted deposits from private as well as from official sources and served somewhat as a national bank, for it had abundant capital. The Temple tax, paid by the Jews until 70 A.D., has been estimated at an annual value of $3,125,000, exclusive of sums brought by pilgrims to Jerusalem.[6]

In Roman times the present industrial relations between East and West were reversed. The Orient was then the great industrial center and exchanged its manufactured products for the raw materials of the less-developed western countries. Palestine's export products, aside from agricultural produce, consisted mainly of textiles. Beisan, formerly Scythopolis, became the center of the linen industry in Roman times. The "fine linen of Beth Shaan" is mentioned in the Talmud, and the export of linen from Scythopolis "all the world over" is cited in a fourth-century source. In 501 A.D. the Emperor Diocletian set maximum prices for all kinds of goods, and the textile products of Beisan headed the list. Exports of fine byssus fabric from Judaea are also mentioned in Greece and Italy. Silk weaving and dyeing is often recorded. In fact, we read of entire villages in the south which were engaged in the latter industry. It has been suggested that the existence of densely populated cities—Kurnub, Khalassa, Ruheeba, Isbeita—in the deserts along the southern route from Aqaba to Gaza can be explained in part by their industrial activities, especially the unraveling of raw silk imported from India and the weaving of mixed silk and linen fabrics. Our word "gauze" comes from Gaza which manufactured and dyed silks and cotton. Purple cloth was produced at Sarafta,

[6] *Ibid.*, pp. 189-193.

Caesarea and even in inland towns, such as Neapolis and
Lydda. Cotton growing and weaving in Palestine are mentioned by St. Gregory of Tours. He says, "Near Jericho, there
are small trees bearing wool (cotton). They grow a fruit,
which has a hard exterior rind, but is full of wool inside.
Even today, such wools are shown; we admired their white
color and fine texture." [7]

Palestine held the monopoly for balsam, a product found
nowhere else in the Roman Empire except in the district of
Jericho. It was a lucrative source of revenue and in the second century of our era brought double its weight in silver.
It was much too valuable to be left in private hands and
became successively the exclusive property of Cleopatra and
of Herod. During the war of 70 A.D., according to Pliny,
there was a battle for every bush. The Jews tried to destroy
the groves to prevent their falling into the hands of the
Romans. This appears to be an ancient version of our modern "scorched earth" policy.

Certain of the country's natural resources served as a
basis for its industrial development. At the southern end of
the Sea of Galilee lies the town of Majdal. In early days, it
bore the name of Tarichaea, which is the Greek for "Pickle-
town," or literally, "the place of salting fish." It was here
that fish were caught in the lake, salted, dried and packed in
jars, to be exported as far as Spain and Rome. Grant maintains that there is no evidence of artificial fish hatcheries
and concludes, therefore, that the Sea of Galilee, with its
ever-fresh flowing water, excellent drainage and suitable
temperature, was rich in natural fisheries. [8] Fish were also

[7] Ibid., p. 191.
[8] F. C. Grant, The Economic Background of the Gospels. New York, 1926.

caught in the Jordan and along the Mediterranean Sea coast.

The ancients had not the scientific knowledge to exploit the vast chemical resources abounding in the Dead Sea. The only two products they utilized were salt and asphalt. The latter was collected from the surface of the sea and exported to Egypt in large quantities where it was used from very early times in the embalming operations preceding mummification. It was also employed in the pitching of ship bottoms.

Living on world sea lanes, and with the example before them of their seafaring neighbors, the Phoenicians, the early Palestinians were also engaged in shipbuilding and transport. Ancient Jewish literature contains numerous references to Jewish seafaring. Technical and nautical terms abound. Jewish sailors were organized into societies which had social aims as well as commercial and professional interests. Josephus Flavius relates that in Tiberias the sailors had a guild of their own. Talmudic sources give us some information about the mutual insurance of ships, practiced by Jewish societies about 1600 years before Lloyd's began to operate.[9]

Thus far, we have examined the story of trade and industry as recorded by ancient writers and historians who have reconstructed the past. But trade was not the real basis of Palestine's ancient prosperity. The needs of the people were simple and luxury imports had little influence on the masses. On the whole, civilization in ancient times was essentially agricultural. Division of labor had not become nearly so complex as it is today. The majority of the population were

[9] Dr. R. Patai, "Jewish Seafaring in Ancient Times," *Jewish Quarterly Review*, July, 1941.

occupied in producing food. One of the best evidences of this is the uniform distribution of the thousands of village sites which cover the landscape like polka dots. Villages served as market places for surrounding farms, and had to be within easy reach of them by transportation on donkey-back or on foot. This limiting factor would tend to create village centers at practically uniform distances.

In addition to supplying local needs, Palestine agriculture in by-gone days produced four main articles for export, one from each of the main divisions of the country: oil from the hills, wine from the plain, dates from the Jordan Valley and grains from Trans-Jordan. The olive, vine and fig have been described as Judaea's staff of life. They "not only sustained her inhabitants, but by their surplus supplied them with the means of exchange for foods in which their own land was lacking." [10]

Trans-Jordan grew considerable surplus wheat for export, and Palestine produced sufficient wheat and barley for local use. The fruit crop was abundant. The produce of the fields in good years seems to have been sufficient to enable the population to bear the heavy taxes imposed upon them by foreign rulers. [11]

F. C. Grant, who is extremely conservative in his evaluation of Palestine's economy, claims that it was self-supporting in the first century. He states that its hills were as richly forested as those of early Greece and England, and that its valleys were fertile enough to produce crops of grains, vines, figs and olives. The breeding of sheep and goats is known

[10] Adolph Buechler, *Economic Conditions of Judaea After the Destruction of the Second Temple.* London, 1912.

[11] F. C. Grant, *The Economic Background of the Gospels.* New York, 1926.

to have been a contributing factor to the economic independence of Palestine.

The amount of building in a country is one evidence of its prosperity, for only if farmers produce sufficient food for the entire population can many laborers be released for other tasks. Moreover, only prosperous farmers can pay the heavy taxes required by rulers for public works. Herod, who collected taxes on a grand scale, spent money lavishly to satisfy his architectural ambitions. He restored old sites and built new cities. These projects served economic as well as strategic purposes. The ruins of Roman cities are to be found on both sides of the Jordan and much information concerning civic buildings, temples and public utility works can be gleaned from literary sources.

But for a land conservationist the most astounding story of this period is written not in history books but on the land of Palestine. One of the most telling evidences of great density of population is the number of abandoned village sites that cover a landscape. There are thousands of them in Palestine. As the country was primarily agricultural, these villages must have been supported by the products of the land. We can readily understand why they were abandoned when we examine the farm lands surrounding them. The once well-built rock-walled terraces have fallen into neglect and ruin, and over extensive areas soils have been washed off the slopes to bed rock, and only dregs of the land are left behind in the narrow valleys. Not only the terraces have been destroyed: the ruins of dams, cisterns, aqueducts and irrigation canals show us how the Palestine of the Hebrew, Graeco-Roman and Byzantine periods reached its high degree of prosperity and why the Palestine of later periods became so much less fertile and greatly decreased in population.

CHAPTER V

DARKNESS OVER PALESTINE

· · · · · · · · · · · · · · ·

"Where there is no vision, the people perish."
—PROV. 29:18

FOR many centuries the people of Palestine, like those of the rest of the Near East, have been plunged into degrading poverty by wars, exploitation and soil erosion. Nowhere has the interrelation between the deterioration of a land and the degradation of its people been so clear as in Palestine. From the twelfth to the end of the nineteenth century the land that had once been so famous for its fertility grew progressively more neglected and poverty-stricken.

Small as it is, Palestine has always been a battleground. On the beautiful and fertile Plain of Esdraelon great armies have often wrestled for the control of Palestine, for it is the natural corridor on the road from the Nile to the Euphrates. Here came the Pharaohs, Barak and Sisera, and here Gideon, with his three hundred braves, smote the enemy "who lay along the valley like grasshoppers for multitude." Here Saul and Jonathan fell in defeat. Then came Xerxes, Sennacherib, Alexander the Great, the Romans under Titus, the Arab invaders, the Crusaders under Richard the Lion-Hearted, Saladin, the Turks, Napoleon. Here, last of all, came Allenby during the First World War.

Aside from the constant wars between external powers competing for control of this fertile and important crossroads, Palestine has suffered time and again from infiltration, invasion and raids by desert nomads. This age-old struggle between the shepherd and the farmer has made the Cain and Abel story of the Bible terribly real in the Near East for thousands of years. The economies of grazing and farming have always been antagonistic, and the desert has always produced more people than it can feed. In times of drought, hardy and desperate marauders have often swept out of the desert and ravaged cultivated regions. Sometimes they even supplanted the former population of the cultivated valleys and became farmers themselves, only to be driven out at some later time by a new nomad horde invading from the desert.

The purely agricultural reasons for the decline of Palestine are a significant chapter by themselves. Agriculture in the Near East was first developed in the flat alluvial lands of Mesopotamia and Egypt, by means of irrigation with the muddy river waters. Soon the farmers of those lands were confronted by two difficulties, each leading to decreased productivity. One was the accumulation of salts in the low-lying irrigated areas. When this happened, lands had to be abandoned and farmers moved on to new areas. The other problem was the stoppage of irrigation canals with silt; this became particularly serious in Mesopotamia. During our travels in that country we examined many ancient irrigation canals now abandoned, stretching out herring-bone fashion on either side of the Tigris and Euphrates rivers. They were marked by banks of silt ten to fifty feet high. Evidently, when the task of lifting the silt out of a canal became too difficult, the old canal was abandoned and a new canal dug alongside of it. We examined eleven canals that had been abandoned one after the other as new canals were dug. The twelfth, nearby, was heavily choked with silt. Great numbers of people were required for the task of cleaning out canals, and the Babylonians appear to have used their prisoners of war for this purpose.

Eleven civilizations have risen and fallen in Mesopotamia during the past seven thousand years. When wars or invasions took away the necessary manpower and interrupted the unending task of cleaning silt out of the canals, irrigation waters were cut off from the fields, villages and cities. Masses of people perished, destroyed more effectively than by the slaughter of war. Sometimes centuries elapsed before another empire arose with a population large enough to restore and maintain a system of irrigation canals.

Between the two great valleys of Mesopotamia and the Nile lay the mountainous region of Palestine, Trans-Jordan and the Lebanon. Farming here was dependent on rains from "heaven," and on these slopes tillers of the soil met with problems which had never been encountered in the alluvial plains. As populations increased, forests were cut, the cultivation line was pushed up the slopes and their fertile soils were planted to food crops. Soon afterwards the soil began to erode under the impact of the heavy winter rains. Farmers have always found the establishment of an enduring agriculture on sloping land their most difficult problem. It is a problem that is still with us after seven thousand years of farming.

We have evidence, however, that tillers of the soil in Palestine and Lebanon early found a way to prevent erosion on slope farming. They built low rock walls on the contour against which soil was washed or filled in, to form benches of level or nearly level strips of land. Where these terrace walls have been kept in repair through the centuries, as we found in certain fields near Jerusalem and on the slopes of Lebanon near Beit Eddine,[1] the soils thus protected have been farmed for three or four thousand years with fairly good results.[2] Along with these practices of conserving soil went works for saving water both for irrigation and for storage in the soil.

How this method of erosion control was arrived at has been the subject of much speculation. We examined numbers of ancient terraced areas in Palestine, Trans-Jordan, Lebanon and on the sites of the ancient Phoenician colonies around the shores of the Mediterranean. The method seems first to have been discovered and applied over the limestone hills of Lebanon, from which it spread to the Phoenician colonies throughout the Mediterranean area and to Palestine and Trans-Jordan.[3] As long as terrace walls were maintained from year to year, soils were held in place for continuous cultivation. If, however, terrace walls were neglected and allowed to fall into ruin, the benches of soils formerly held in place by the walls were exposed to more rapid erosion than if terrace walls had never been built. As we shall see, the period of intensive terrace farming in Palestine was replaced by an entirely different and much more primitive

[1] W. C. Lowdermilk, "The Cedars of Lebanon," *Am. Forests,* January, 1941, p. 34.

[2] Nelson Glueck, "Earliest History of Jerash," *Bulletin of the American Schools of Oriental Research,* No. 75, p. 23, 1939.

[3] Paul Vidal de la Blanche, "Les Grandes Agglomérations Humaines," *Annales de Géographie,* 1918, p. 179.

culture brought by the Arab invaders from the grazing lands of the desert.

Decline after Arab Invasion

The decline of Palestine's land and of the people began with the first Arab invasion during the seventh century of our era. However, several centuries were to elapse before the country reached the stage of utter desolation. Although the Nomads destroyed many cities and overran cultivated areas with their destructive herds, much of Palestine seems to have escaped the full effects of this first wave out of the desert. The country still was fertile and many of the native population were allowed to remain and to continue their traditional ways of farming or to carry on their former trades. It was not until the wars of the Crusaders during the twelfth and thirteenth centuries, and the second Arab invasion which drove them out, that Palestine was plunged into its age of darkness.

The Arab traveler, Mukaddasi, who visited Palestine in the tenth century just before the Crusades, describes the country as generally fertile. Following are several excerpts from his writings:

The mountains of Busra (the Hauran) were covered with vine-yards. Tiberias was renowned for its crops and around its shores were villages and palm trees, and on its surface were boats and the water swarmed with fish. . . . In upper Galilee are many fine villages and here are grown grapes and other fruits and olives and there are many springs. The rainfall waters its fields. . . . Beisan abounds in palm-trees and from this place comes all the rice consumed in the provinces of the Jordan and Palestine. . . . Ramleh, Palestine's capital, is well built and has plentiful water, abundant fruits and prosperous commerce; it is easy to make a living here. Its fruits are the most luscious and its bread is the whitest. . . . It has spacious houses and good roads. The plain on which it stands needs no irrigation for the fields are naturally rich and fruitful. . . .

. . . Jerusalem has clean markets, enormous grapes and incomparable quinces. . . . No place can excel Hebron in its beauty, its orchards of apples, its ground bearing grapes and its fine quality fruits, a great part of which go to Egypt. . . . At Beit Jebrin, are marble quarries. It is a land of riches and plenty, possessing fine domains. . . . The population, however, is on the decrease. . . . The lands of Caesarea are excellent and its fruits delicious. . . . Nablus (Shechem) abounds in olive trees and there are some remarkable mills there. . . . Jericho grows much indigo and many palms, bananas, dates, and flowers of fragrant odor. . . . Trade from Palestine includes olives, dried figs, raisins, and the carob fruit, also stuffs of mixed cotton and silk.[4]

While the country still retained much of its ancient fertility and prosperity, there were also harbingers of decline. Ishtakhri, a commercial traveler in the tenth century, wrote: "The district of Moab is extremely fertile and rich, only the Bedouins have the upper hand here, and so ruin all." [5]

The Crusaders, from 1096 to 1291, did not bring happiness to the people of Palestine. The record they left is one of envy, hatred, malice, vice and diseases, all rendered the more repulsive because the Crusades were carried on in the name of Christianity. The Crusaders fastened onto Palestine the heavy yoke of feudalism as they brought it from Europe, and for eighty-eight turbulent years imposed a feudal kingdom

[4] C. Mukaddasi, *Description of Syria, including Palestine,* translated by Guy Le Strange. London, 1886.
[5] Guy Le Strange, *Palestine Under the Moslems,* p. 35. London, 1890.

on the country. Palestine was divided into districts owned and ruled with a high hand by feudal lords, societies and religious organizations. Farmers who worked the soil, or Bedouins who herded the sheep, belonged to the land, and went with its sale. There is on record a certificate dated 1179 and made out by a Crusader from Shechem which reads: "The sale of all my Bedouins, and all the others who live in tents in any place whatsoever, with all their families and possessions." [6]

While ruinous as a whole, the rule of the Crusaders had a few advantages. One of them was foreign trade that stimulated the growth on the coastal plain of crops such as indigo, sugar cane and wine, which found ready markets abroad. The ports of Palestine became crowded with European boats and the coastal cities flourished. The interior, however, was laid waste and deserted through strife and wars and the plundering of Bedouins. This was particularly true during the last period of the Crusaders' rule. Nachmanides, who came to Palestine in 1267, reports that Jerusalem was deserted and laid waste, and that Judaea was more destitute than Galilee.

In the declining period of the Crusaders' rule, the exploitation of the tillers of the soil became continually worse, while plague and epidemics killed off much of the population. Trade decreased and cities shrank in size and importance. The State owned the soil and overlords squeezed all they could from peasants. As a result, the peasants (fellaheen) neglected the land. Their chief interest was to get something from the next harvest, and to avoid tax collectors and plun-

[6] K. G. Preston, *Rural Conditions in the Kingdom of Jerusalem During the Twelfth and Thirteenth Centuries*. Philadelphia, 1903. 59 pp.

dering Moslem nomads. Sometimes fellaheen were forced to abandon farming and to become semi-nomads.

During the Crusaders' rule the "enchanted forest" in the Valley of Sharon was cut down. But great forests of oaks in the region of Caesarea evidently came back to an extent, and in 1737[7] Pococke noticed great quantities of wood along the shore ready to be shipped to Egypt. M. Paultre, who was with Napoleon's army on its march from Jaffa to Acre, reported the existence of these trees.[8] This famous lowland forest is entirely gone today, but traces of it are now found when clearing land for modern orange groves, and in a few remaining "sacred trees."[9]

After the expulsion of the Crusaders and a new invasion by Arab nomads, the decline of Palestine proceeded at an accelerated pace. The economic situation deteriorated. Taxes became unbearable, and the oppression of the fellah most cruel. Lawlessness and insecurity prevailed. The rulers of the land punished villages by cutting their trees and sometimes by totally destroying them. At this time fellaheen hid in the hills; it was difficult to find land to cultivate there, but at least they were somewhat protected from the attacks of tax collectors and savage Bedouins. There was no established government and anarchy prevailed everywhere. Desert Arabs poured into the land and to the very gates of Jerusalem, stealing and plundering on the roads throughout the country. The country became a desert land with no one to till the soil. Caesarea was utterly destroyed; Jericho became merely

[7] Richard Pococke, "Description of the East," in G. Pinkerton ed., a general collection of *Voyages and Travel*, Vol. 10, 1811, p. 445.

[8] Jacob De Haas, *History of Palestine.* New York, 1934.

[9] A. Eig, "Essay on Palestinian Forests," *Sonderabdruck aus den Beiheften zum Botanisches Centralblatt*, Bd. L I. 1933, pp. 226, 254-263.

a collection of huts; Lydda was but a poor village and much of Acre was in ruins.[10] In this state of chaos, terrace walls and works of water conservation were allowed to fall into ruin.

The decay of Palestine reached its darkest stage in the four hundred years of Turkish rule, from 1517 to 1918. The old Turkish regime was very different from the rejuvenated and progressive Turkey of today. In Palestine, as in the rest of the old Turkish Empire, appallingly high taxes were levied on the tillers of the land. Trade was poor and prices exceedingly high. The privilege of collecting taxes was sold to the highest bidders among rich or powerful individuals belonging to entrenched families or strong political cliques. These in turn farmed out the taxes to lesser agents who unmercifully exploited the peasantry. The government sought maximum revenue but gave little in return.

Although many of the fellaheen were descendants of the ancient farmers of terrace agriculture, under these oppressive conditions they lost all interest in careful, far-sighted cultivation. Possessing none but the most primitive tools, they were, moreover, frequently forced to loan their farm animals for military use during part of the year. At the same time Bedouin nomads robbed the fellaheen of all produce they could lay hands on. Often the protection paid by farmers to Bedouins to prevent total plunder equaled one-third of the government taxes,[11] and tax collectors left little of the balance. After threshing time, troops of police would surround villages to collect the taxes. One of the worst curses of the Turkish regime was the tax imposed on every tree,

[10] De Haas, *op. cit.*, p. 321.
[11] *Ibid.*, p. 364.

whether woodland, forest or orchard, and on each vine as well. Many fellaheen found it more convenient to cut or dig up their trees rather than pay these heavy taxes. This destructive system of taxation accounts in part for the treeless condition of Palestine at the end of the nineteenth century. Sometimes an entire village was abandoned and the fellaheen became Bedouins or semi-nomads. The peasants had to sell all their produce and live on the coarsest food. The poorest European day laborer was better off than the Palestine farmer at that time.

Moreover, the old Turkish government became too weak to hold local sheikhs in check when they quarreled with each other. Sheikhs exerting great influence in the various districts into which Palestine was divided were allowed to interfere with the rule of the local Turkish officials, who in turn did not present a picture of virtue. Petty warfare between neighboring sheikhs brought destruction to lands, crops and people. Maundrell, an English chaplain at Aleppo in 1697, said: "At Acre there are a few poor cottages—the surrounding country is a vast and spacious ruin, and Jericho is a poor nasty village."[12] "The country had a bedraggled appearance, most villages were contemptible and except for the forest in Sharon and some fruit orchards, the land was bare of trees," reports George Sandys, who visited Palestine in 1610.

There were a few exceptional periods when strong and far-sighted local rulers established more security in the ravaged country and thereby alleviated the burden of the fellaheen. Thus encouraged, tillers of the soil and town dwellers were stimulated to greater enterprise, which quickly

[12] Henry Maundrell, *The Journal from Aleppo to Jerusalem, 1691*, Bohn's edition. London, 1848.

resulted in comparative prosperity. However, such periods of respite did not last long. The Turkish government, suspecting such local governments of separatist tendencies, quickly liquidated them.

In 1798 Napoleon conquered Egypt and marched into Palestine, but was defeated at Acre, owing to the assistance given the Turks by the British fleet. Napoleon had some far-reaching plans for the Near East, one of which was the restoration of Palestine to the Jews. But his short-lived invasion had little influence except to bring further destruction upon the country. During his campaign more forests were destroyed.

By the middle of the nineteenth century, the conditions of the population and the land reached their lowest ebb. Insecurity prevailed and Bedouins plundered farms and robbed travelers and caravans on the roads. Petty officials cruelly exploited the defenseless fellaheen. The grazing culture of the desert Arab completely replaced the refined ancient agriculture. Herds of goats overran the country, leaving the land defenseless before runoff from the heavy winter rains which year after year eroded soils into the valleys and out to sea.

During these years of decline, Palestine had been gradually depopulated. The lowest point was reached about 1850 when the total population was below 200,000. Nomads held full sway over all the Negeb and made pillaging expeditions far into Palestine proper. The Tayaha tribe, oldest and most important in the Negeb, gathered hundreds of warriors once a year to plunder the surrounding country.[13]

[13] Eliahu Epstein, "The Bedouin of the Negeb," *Palestine and Middle East Magazine,* 1937, pp. 524-530.

As late as 1877, Colonel C. R. Conder said that Palestine was empty and the population not large enough to till the land. He suggested that the land could support ten times as many people as it then had.[14] United States Consul T. G. Wilson, in his consular report of October, 1881, said the Bedouins had driven the population to the hills, and the plains were neglected.[15] Dean Stanley spoke of the desolation of the country, stating that "the Arab population is incompetent and lethargic and there is dismal want among the Jews.[16] Russell, in his *Palestine, the Holy Land,* ascribed the condition of the land to the barbarism in which the great mass of the population was immersed. He noted vine stalks exceeding a foot in diameter and said that Galilee would be a paradise if an industrious people lived there.[17]

Around Ramleh near Jaffa, during the middle of the nineteenth century, there were no more than a hundred miserable families living in huts where they quartered their cattle with them in winter. In that district there were superior olive orchards, but these were allowed to perish through old age and the ravages of contending factions.[18] Often revenge was taken on an enemy by cutting his trees at night. Wells were filled with debris and cisterns decayed and cracked for want of care. The ruins of huge vaulted reservoirs prove that the ancient town of Ramleh had probably been more than four miles in circumference.

According to Russell, too, Jericho, once famous for its

[14] Jacob De Haas, *The History of Palestine*, p. 407. New York, 1934.
[15] U.S. Department of State, U.S. Consular Reports, October, 1881, p. 514.
[16] De Haas, *op. cit.*, p. 407. See A. P. Stanley *Sinai and Palestine*, p. 120. 1857.
[17] M. Russell, *Palestine, the Holy Land*, p. 31. New York, 1832.
[18] *Ibid.*, p. 1205.

[77]

precious balsam (*Commiphora opobalsamum*), on which Palestine had gained much wealth, was treeless in 1850. No palm or balsam trees were to be seen near this deserted town and this complete desolation is attributed by him to the people, rather than to any perceptible change in climate.

The inhabitants are described by Carnet in *Recollections of Travel in the East* as "barbarous, inactive and despondent, and in constant fear that the fruits of their labors may suddenly be reaped by any of the oppressing or marauding chiefs who have the strongest hand at the moment." Taxation forced these poor peasants to sell their few animals, sheep, goats and fowl, as well as the chief part of their produce and live on what poor food remained.[19]

The Turkish government farmed out tax collection to the highest bidders who were usually rich effendis (landowners) living in the cities. According to law, the maximum tax consisted of 15 per cent in kind, but the tax-collecting effendi evaded this by assessing the produce at more than the real value and insisting on cash. Compulsory military service took his grown sons from the peasant and farmers were eager to buy the release of their sons with money. The city-dwelling effendi loaned the cash, but henceforth the farmer became the effendi's debt-slave for life, for he could not pay as much as the interest. When the harvest was over, each working family was given a bag filled with grain, which they had to thresh for themselves. This meager quantity of grain had to suffice the family for an entire year, aside from the products of a few fowls and animals.

The low cultural level of the fellaheen suited the rich effendi. No schools were provided and no doctors were avail-

[19] John Carnet, *Recollections of Travel in the East*. London, 1830.

able for the country folk. Peasants suffered severely from malaria and trachoma, an eye disease that was very prevalent in Palestine then and still causes much blindness. Village and home life was wretched. Carnet gives a vivid description of an Arab hamlet among the rocks on a hill opposite Jerusalem which is little different from villages we saw in Trans-Jordan in 1939. Some dwellings were only caves out of the rock. Others were miserable hovels, with flat roofs. No trees or spots of green were to be seen on the gray landscape. The inhabitants were "squalid," standing or sitting and "listlessly gazing into the valley beneath." Inside were but a few wooden utensils and some coarse mats to sleep on.[20]

The Problem Today

The results of Palestine's period of neglect are still very apparent in the land. Striking examples of devastation and progress may often be seen side by side. Outside Hadera, one of the older Jewish settlements, we saw a group of black and brown goathair tents belonging to semi-nomads, near a patch cultivation of grain that could not have yielded more than five bushels to the acre. This was the most primitive sort of life, without a settled abode, without sanitation, without education or any of the cultural advantages of modern civilization. Just across the road was a modern Jewish dwelling, surrounded by a well-tilled orange orchard, its waxy deep-green leaves dotted with white fragrant blossoms and illumined with ripening fruit. A farmer was cultivating the land between the trees with a tractor drawing a many-shoveled cultivator. Here, in sharp contrast, were the two

[20] John Carnet, *op. cit.*, p. 152.

types of agriculture and modes of living that have separated grazing from farming since the time of Cain and Abel.

Under ordinary circumstances, centuries would have to pass before the more primitive people could reach an equal stage with their advanced neighbors. However, the stimulation given by the example of the Jewish colonies may very possibly cut short this transition period. Otherwise we cannot expect much progress. Mandate officials had little success in their attempts to get the Arabs to plant trees, even though the seedlings were free and assistance in planting them was given. Major Jarvis, who spent many years in Sinai and southern Palestine, writes: "As for the Arab, he is so thoroughly satisfied with the system employed by his ancestors a thousand years ago that it is absolutely hopeless to try to get him interested in a new vegetable or seed or any change that savors of modernity."[21] In other words, progress is impeded by the temperament of the people and their fatalistic philosophy of "*Mectub*,"—the belief that whatever happens is predestined.

Along with the need for improved techniques in the use of land and the development of industries, Palestine, like the rest of the Near East, must have a larger population if it is to be restored to full activity and prosperity. There is a close correlation between the decline in population and the wastage of hill lands in Palestine during the past thousand years. A survey made by P. L. O. Guy, Director of the British School of Archeology in Jerusalem, in the catchment of the Wadi Musrara draining the western slopes from Jerusalem to Tel Aviv, disclosed that out of a total of 293 village sites found in that region, 193 are now abandoned. In the plain, 32 sites are

[21] C. S. Jarvis, *Yesterday and Today in Sinai*, p. 265. London, 1931.

[80]

occupied and 4 abandoned. In the foothills, 31 are occupied and 65 abandoned; in the mountains (975 feet or over), 37 are occupied and 124 are abandoned. Most of the abandoned sites were occupied during the Byzantine period and seem to have been deserted following the Moslem invasion. We examined abandoned sites near other wadies also, and found that there, too, the soil was impoverished or eroded away over extensive areas.

The lands of Palestine, except the 14 per cent of arable land which has been acquired by Jewish settlers and a few progressive Arab estates, are a sorry commentary on man's exploitation of a once fruitful country. The barren rocky hills have been accepted by the natives and the casual observer as a normal condition impossible of improvement. The gullies, gouging out the remaining soils in the valleys, the destructive goats of the fellah and Bedouin, the poverty and low standards of living of the population, are considered inevitable by those who fail to read the tragic story of land wastage and exploitation inscribed deeply in the landscape.

Some students of Palestine have advanced the theory that an adverse change of climate must account for the impoverishment of the land and decadence of its people. We found ever-increasing evidence in our extensive study that the decline of the Near East was due not to adverse change of climate but to man-made devastation and neglect.

The theory of climatic change or pulsations of climate was advanced by Huntington,[22] who claimed that this change was mainly expressed in the decrease of rainfall, whereas tem-

[22] Ellsworth Huntington, and S. S. Visher, *Climatic Changes, Their Nature and Causes*, p. 95. New Haven, 1922.

perature differs little from that prevailing at the height of Palestine's ancient civilization.

Scientific evidence indicates that the climate of the earth in general has become warmer since the Ice Age but not necessarily drier. It is indeed possible that annual rainfall since the Ice Age may have actually increased. For if the climate has become warmer evaporation from the sea should be greater.

I met with a similar theory of climatic change during my studies of land use and erosion in China. Baron von Richthofen, the geologist, believed that adverse climatic change was responsible for the decadence of North China, where vast regions, once fertile, had become denuded and devastated. But in the course of my agricultural explorations there I found temple forests, protected by Buddhist priests, where tree stands were reproducing themselves naturally without irrigation or plantings, just as they had done prior to the time when men, by pushing cultivation up the slopes, exposed the soils and permitted the forces of soil erosion to complete their destructive work. No appreciable soil erosion took place within these forests because the soil surface was covered with forest litter. But surrounding these temple forests in the great loessial soil region, the whole country was riddled with enormous gullies still head cutting and eating back into the landscape.

The same situation prevails in Palestine. Had the amount of annual rainfall decreased since Roman times, the levels of the ground water in springs would be lowered. With Drs. Glueck and Harper, the archaeologists, we examined several springs in Trans-Jordan walled with heavy masonry built in Roman times. In these springs the masonry and apertures fitted present water levels and quantity. This indicates that ground water supplies have not changed since Roman times. In Amman, the ancient Philadelphia, where the supply of water seemed to be under the Roman level, the difference immediately disappeared after a thorough cleaning of the spring.

That temperature in modern Palestine are about the same as those prevailing in antiquity is shown by the fact that there is no appreciable difference in the distribution of the date palm, which is generally recognized as a sensitive key to climatic change.

We found further evidence of the constancy of the climate in our study of the remnants of cedar groves in Lebanon. About three hundred years ago a cedar grove consisting of forty-four scattered veteran trees was made a sacred place and a Maronite church was built among them. A stone wall around the grove protected it from the ubiquitous goats. This protection sufficed to permit seed from the veteran trees to germinate and grow up into sizable trees. Competing for light, these younger trees grew up into splendid straight stems. The inevitable conclusion is that the present climate in Lebanon is no less favorable than the ancient climate to the growth of the famous cedars.

Similarly, in many cases the young trees and forests in the Jewish colonies of Palestine thrive on what appear to be bare rocky slopes. The tree roots search out the rich soils left in the solution pockets of the limestone country rock. This growth would not be possible if the climate of Palestine were drier now than in ancient times.

While the rainfall is essentially the same, its effectiveness has been decreased by the desolate condition of the country.

Soil erosion increases the runoff of rain water, and the absence of forests reduces fog or cloud-drip from the trees. Thus, man-made devastation and resulting soil erosion may make the same supply of water less effective.

Our general conclusion is that the gradual deterioration of Palestine has been due, not to climatic change, but to man-made causes, and that productivity depends today as in ancient times upon the care and protection given the land. If men in Palestine will treat their land with the full care it deserves, and if a far-sighted government will assist them in their efforts toward reclamation and conservation, the country will again become a "land wherein thou shalt eat bread without scarcity."

CHAPTER VI

NEW FARMERS IN A NEGLECTED LAND

. . . .

*"The wilderness and the solitary place shall be
glad for them."*
—ISAIAH 35:1.

THE modern Jewish resettlement of Palestine began in 1882
after a wave of pogroms in Russia, which occurred to a large
extent with the knowledge and approval of the Czarist government of that time. Coming after a long period of comparative calm, these bloody attacks stimulated large-scale
Jewish emigration from Russia and other parts of eastern
Europe. While most immigrants went to the United States,
some of them settled in Palestine with the idea of reviving
the land of their ancestors and making it again into a Jewish
National Home. They felt that only thus could they provide
a realistic and permanent solution for the Jewish problem.
While the Jewish immigrants to the United States fitted themselves into an existing economic structure, often contributing
to its further development, the newcomers in Palestine became pioneers, building their own economy in a backward
and neglected land. They realized that the foundation for
an enduring state must be anchored in agriculture, and to do
this required improvement of the land regardless of difficulties, toil and dangers.

These pioneers in Palestine found the country denuded of trees and depleted of its natural fertility. Since they were all originally city dwellers, they had no agricultural experience and at first imitated the methods by which peasants worked the soil of the rich Ukraine and the other eastern European lands they had left. After superficial plowing, they sowed wheat in the sandy soil of the Maritime Plain of Judaea and they were disappointed by the failure of the crop. They bought emaciated little cows from their Arab neighbors and were dismayed at the low yields of milk. They planted trees on soil underlaid with a clay pan called "*nazas*" and were exasperated when the clay layer during the hot summers became as hard as concrete, cracking and killing the tender roots of their trees.

These experiences quickly taught the first settlers that satisfactory crops could be grown only after considerable improvement of the soil. They were eager to learn and to experiment—that was the positive side of their lack of agricultural tradition. Hence they sought advice everywhere, turning to the agricultural experts sent to Palestine by their benefactor, Baron Edmond de Rothschild, and frequently visiting the few German colonies which the Templars had established on the Maritime Plain about a decade before. At the same time they studied the method of the Arab fellaheen, and sought to learn from the experience the latter had accumulated during centuries of primitive trial and error. Years later, experts at the Agricultural Experiment Station in Rehovoth were to find in that experience points of departure for the improvement of Palestinian agriculture.

From the beginning it was apparent to the Jewish settlers that they must do far better than the fellaheen had done.

They knew that the success of their colonization depended on their finding the right answers to the many agricultural problems that confronted them, and they concentrated on these problems with all the keenness of minds sharpened by generations of intellectual training. In less than a generation they greatly improved their methods of farming. Deep plowing and the eradication of weeds preceded the sowing of every crop. Heaps of manure that had accumulated for centuries around Arab villages were bought to fertilize the lands of the Jewish colonies. Citrus plantations were started only after soil analysis and each grove was surrounded by a wall to keep out goats, the "black-eared locusts" of the Near East. Soil was improved by growing leguminous crops which were plowed under to increase nitrogen and the much needed organic content of impoverished land. The crops of the Jewish villages soon exceeded those of the near-by Arab villages.

Much of this advance was due to the influence of Mikveh Israel, the agricultural school established in 1870 near Jaffa, where young people were given technical and practical training in farming. I was shown over this interesting institution where students are required to devote one-half day to their studies and the other half to practical work on the big school farm. Their work is rotated through all activities of the farm enterprise.

In the course of founding new settlements, Jews bought neglected tracts of land that had to be reclaimed by various means. In many areas swamps had to be drained on a large scale to banish pestilential malaria. The settlers introduced eucalyptus trees from Australia and planted them on swampy areas to dry up excess moisture and destroy the breeding pools of mosquitoes. The Arabs promptly called them "Jew trees." These Palestinian eucalyptus trees resemble those of Southern California. Jewish pioneers also planted pine trees and castor oil plants on the rolling sand dunes near the seashore to control their movement and prevent them from advancing farther over farm lands and villages. Later on, many of these dunes were leveled and planted to grapes and citrus fruits with surprisingly bountiful results. Valley lands on which torrents from the hills swept gravel and debris had to be cleared for cultivation. The stones were built into the diversion dikes and dams to protect cultivated lands from further damage by floods. These measures remind us of how cultivated outwash fans in Southern California must also be protected from torrential winter rains.

Dry Farming and Irrigation

Jewish pioneers at first concentrated on dry cereal farming. But this type of farming is not fitted to the climate and soil conditions of Palestine; it offered a fair return only in the valleys of Galilee with their richer soil and more abundant rainfall. After initial failures, the pioneers of Judaea and Samaria accordingly turned their attention to vineyards and other suitable tree cultures, in which they achieved much better results. However, they soon encountered difficulties in marketing these products, and suffered from the vicissitudes attendant upon any monoculture. They were further handicapped by the lack of auxiliary farming—poultry and vegetable raising for their own consumption.

As time went on, this one-crop system gave way to mixed farming which made each family or group self-sufficient on the land and independent of the hazards of over-commercialized agriculture. Many of the younger settlers who followed the veterans of the 1880's came to the conclusion that they could not redeem the Promised Land with hired labor but that they must work with their own hands. A few years before World War I, the Zionist Organization, later organized as the Jewish Agency, began to encourage settlements based on co-operation and excluding the employment of hired labor, whether Jewish or Arab. Most of the many scores of settlements founded since the last war belong to this type and are financed by two Zionist funds. The first of these is the forty-one-year-old Jewish National Fund which purchases land as the national property of the Jewish people, and leases it afterwards on long-term contracts to settlers willing to abide by definite stipulations, including the tilling of the soil by their own labor. The second and newer of the funds, the twenty-three-year-old Keren Hayesod (Foundation Fund), was given the task of financing the development of the villages by the settlers who in most cases did not possess money of their own.

Jewish colonization in Palestine was confronted with the problem of establishing a type of agriculture that would support a comparatively advanced European standard of living in the midst of the primitive subsistence farming and the low standards of living prevailing in the East. The products of Jewish settlements have had to compete with the low-priced crops of Egypt and Syria where labor, especially that of women and children, is sadly exploited. Furthermore, the Jewish colonizing agencies have had to make full-fledged farmers out of people whose background had been urban for many generations.

A very important technical problem faced by Jewish agriculturalists is that of irrigation. In most of Palestine there is enough rain to make possible the growth of cereals and certain vegetables and fruits (grapes, figs, olives, almonds, etc.); the yields are, however, comparatively low and always subject to seasonal hazards and variations. It early became apparent to the pioneers that they needed a supplementary supply of water for domestic use and for the vegetable plots around their homes. Using the primitive methods of the native population, they began to dig wells. The yield of these wells was unsatisfactory until years later, when American well-boring machines were brought to Palestine. During the fifteen years between 1924 and 1938, the Zionist colonizing agencies dug 548 wells and installed several canal systems to spread water from springs and streams.

At the present time in Palestine over 95,000 acres of land are under irrigation. Most of them lie in the Maritime Plain. They are largely devoted to citrus plantations for which irrigation is necessary even as it is in Southern California. These 95,000 acres are slightly more than an eighth of the area of 741,000 acres considered irrigable by Dr. Strahorn in the soil survey he made for the Joint Palestine Survey Commission in 1926. Moreover, there can be no doubt that the area under irrigation, especially for vegetables and fodder, can be greatly increased beyond Dr. Strahorn's estimate if the Jordan Valley reclamation project suggested in the further course of our book (Chapter XI) is realized after the war.

Livestock and Poultry

The first Jewish settlers found the native cow to be a rundown breed. For centuries, the poor creature had been left to forage for itself during the long dry summers without any supplementary feeding. On the other hand, cattle imported from Europe were badly affected by the climate of Palestine. Accordingly, the Jewish settlers began to cross Holsteins from Holland and Jerseys from England with the Syrian and Lebanese breeds which are superior to the native Palestinian cow. They managed to develop a strain of milk cow that could survive and prosper under Palestine conditions and yield a satisfactory quantity of milk. The Arab cow gives about 800 quarts of milk a year but the cows in Jewish settlements average 3500 quarts, while some prize specimens among them give more than 5000 quarts yearly. While we were in Palestine, negotiations were completed to export this superior breed of acclimatized cow to India.

Today Palestine's Jewish settlements supply ever increasing quantities of liquid milk, and produce much butter, cream, cheese and other milk products. The sales of Tnuvah, the largest marketing co-operative of Jewish farmers, show that in 1924-25, 1,131,672 liters of milk were sold. In 1935-36, the sales reached 18,074,557 liters, and the estimate for 1941-42 is 26,000,000 liters.[1]

Jewish farmers are rapidly increasing their herds of sheep and using them for meat as well as for milk and wool. A still larger production of meat will become possible with the introduction of more scientific grazing. Recently, the native fat-tail sheep has been crossed with the caracul sheep of Central Asia. This has resulted in a notable new animal that not only has the qualities of the fat-tail sheep which stores up fat for the lean and dry summers, but carries a coating of fur that is said to compete satisfactorily with Russian and Persian caracul skins in the London fur market.

[1] A liter is 1.06 quarts.

The native Arab hen is a scrawny fowl which lays an average of 70 small eggs per year. Jewish settlers have introduced the leghorn hen and crossed it with native breeds. The resultant new strain is heavier and lays an average of 150 large eggs per year. This new breed is now found all over Palestine.

During our travels in Palestine we visited a modern poultry colony. There we found a group of refugees, all doctors of medicine or law, who had escaped with some money and had turned their trained minds to the task of producing more, bigger and better eggs. They also specialized in growing baby chicks which they sold to Jews and Arabs alike. Their methods of work were as modern and scientific as any we had seen in other countries.

In general, the use of scientific agricultural methods on irrigated land has produced noteworthy results in Palestine. An irrigated acre of beets used as concentrated food for milk cows yields forty and sometimes even fifty tons. Irrigated clover yields eight harvests a year. Similar results are obtained from alfalfa introduced from the United States. Clover and alfalfa fodder are now the mainstay of Palestine's dairy farms, for natural pasture is available only during the six rainy months and must be supplemented by hay and feed crops during the long dry season. Experience has shown that fodder grown on a half-acre of irrigated land is sufficient to maintain a cow during the dry months.

While the results in cereal crops are not as striking as in dairy products, the yields of wheat, barley and corn on Jewish farms are nearly twice as great as on Arab farms. Thus 25 bushels of wheat per acre is the average yield in the Jewish villages of the Esdraelon Valley, as compared to about 13 bushels in the near-by Arab villages.

Vegetable Farming

Some twenty years ago the choice of vegetables in Palestine was very limited, and the quantity far from adequate for the population, but since the discovery of underground water resources in several districts, vegetable production has increased enormously. The Jewish villages have introduced many new varieties and improved the quality of the native vegetables. One now can buy vegetables every month in the year, and they are recognized as an essential part of rational crop rotation in Palestine. Summer vegetables are grown during the winter in the tropical Jordan Valley. If we include melons and watermelons among vegetables, the vegetable area of Palestine exceeds a half-million dunams, or 125,000 acres, which is about 7 per cent of its total cultivated area. Most of the area under vegetable crops relies exclusively on rainfall, and only about a third of it is under artificial irrigation. Excluding melons, the vegetable area of Palestine rose from 12,563 dunams in 1923 to 142,364 dunams in 1938 and by now exceeds 300,000 dunams, or 75,000 acres. Next after melons, which occupy about a third of the total vegetable area of Palestine, are tomatoes, whose rapid development in Palestine matches that in the United States. Vegetables, especially under irrigation, are destined to take a still larger place in Palestine's future agricultural economy.

About ten years ago, Jewish farmers were quite pessimistic about producing a large potato crop in Palestine. The difficulty was not in growing the potatoes but in storing them during the hot months of the year. This problem has now been solved by providing the stored potatoes with enough ventilation to prevent premature sprouting, and the potato crop is now rapidly increasing.

Quite recently Jewish farmers introduced the culture of sugar beets and sugar cane on a commercial scale. A sugar refinery in the eastern part of the Plain of Esdraelon uses beets grown in the cooler western part of the valley and sugar cane raised in the adjoining valley of Beth Shean, which is slightly below sea level and sub-tropical in climate. The seasons for sugar cane and beet sugar are complementary, and the new factory will therefore have the advantage of working through most of the year.

Honey, for which Palestine was famous in ancient times, is an important by-product of its agriculture. Many farmers possess hives of bees which extract their honey from orange and other orchard blossoms, and from the great numbers of field flowers which we found coloring the landscape during spring. The production of honey has greatly increased in recent years as a result of newly introduced scientific methods. In 1936 the amount of honey produced was 900,000 pounds and it is now estimated at 1,200,000 pounds. The potential output is estimated at about 5,000,000 pounds annually in the Jewish agricultural colonies alone.

Development of Citriculture

The most outstanding achievement of Palestine agriculture is the scientific production of citrus fruits. When the Jewish settlers came, they found Palestine producing a high-quality orange, the Shamuti, which they at once began to cultivate and improve. They also introduced grapefruit, which we all agreed was the finest we had ever tasted. The Washington navel and Valencia oranges were also introduced with successful results. Generally speaking, the citrus groves in Palestine closely resemble those of Southern California, and in some ways the Palestinian fruit seemed superior to that of California.

The Arabs quickly adopted the modern methods of the Jews and together the two communities have built up a citrus industry which before this war made Palestine the second largest citrus exporting country in the world. Expansion on a large scale began after Jewish research established improved methods of planting, picking, packing and marketing. Nowhere in the world have I found citrus groves cultivated as intensively and scientifically as in Palestine. In some of the larger groves one finds a complete filing system in which the "case history" of each tree is entered.

The distances between individual trees in Palestine's orange groves are less than in California. Between 190 and 200 trees are usually planted to the acre while in California the average is 100 to 120 trees. This practice is partly explained by higher land costs, and results in larger crops during the sixth to twelfth years of the trees' life. After that period a mature grove in California equals the yields of Palestine and even excels them. In Palestine there is no need for the expensive oil stoves which are used in California to protect the trees and crops in freezing weather.

Eighty-five per cent of the area under citrus is devoted to oranges. Second place is taken by grapefruit, which is a comparative newcomer in Palestine. Lemons are third. The quantities of citrus exported during each of the last three years before the war are given in the accompanying table.[2]

As the domestic market is limited by the small size of the country, about 85 per cent of the citrus crop was sent abroad

[2] The box used in Palestine contains from 100 to 160 oranges, and from 95 to 125 grapefruit, depending on size.

[95]

CITRUS EXPORTS OF PALESTINE (in boxes)

Year	Number of Boxes
1936-37	10,795,894
1937-38	11,444,408
1938-39	15,264,776

in normal years. Most of it went to Great Britain. While the countries of continental Europe have become increasingly vitamin-conscious in the last generation, their purchase of citrus was restricted by their governments, which designated citrus fruits as "luxury imports" and discouraged their consumption by imposing high tariffs and quotas.

For a time, investment in citrus groves was both popular and very profitable. But since 1933, poor seasons, lack of tariff protection, overproduction, glutted markets, difficult transportation and later the war checked the further expansion of citriculture.

In addition, Palestine's citrus industry suffers from a short harvesting season. Until recently, nearly all oranges were of one variety—"Shamuti." Their harvest begins at the end of December and ends in the first week of April. Efforts are being made to prolong the season by planting varieties that ripen either earlier or later. If these efforts succeed, the selling season will be prolonged by two or three months and the prospects of marketing Palestine's citrus crop correspondingly improved.

When the war interrupted normal sea traffic, the depression in Palestine's citrus belt became acute. Nearly all the 1940-41 and 1941-42 crop was left to rot on the trees or was buried in the ground as a kind of "ersatz" fertilizer. This condition has since been somewhat alleviated by increased domestic consumption, purchases by the British and Allied armies stationed in the Near and Middle East and expansion of the canning industry. Furthermore, Palestinian scientists are now utilizing citrus for the production of alcohol, vitamin pills (ascorbic acid) and pectin and have developed a way to use quantities of orange pulp for feeding of dairy cows.

The problem of Palestine's citriculture rests ultimately with the future of world markets. In a period of peace and prosperity increased consumption of citrus fruits and a relaxation of trade restrictions are probable. The demand for vitamin-containing fruits after the malnutrition of the war years will presumably insure enlarged markets in Europe for the conveniently located Palestinian citrus crops. Despite these encouraging prospects, the leaders of Palestine's agriculture will act wisely if they avoid the vicissitudes inherent in monoculture by continuing the trend toward the further development of mixed farming.

Other Fruit Cultures

The wine industry, on which the earliest Jewish pioneers laid great hopes, does not now occupy an important place in Palestine's rural economy. The large modern wine cellars built by Baron de Rothschild in the older villages of Palestine are not generally utilized to their full capacity. Palestine produces some excellent wines, especially of the Spanish and Portuguese brands (muscat, Malaga, port), but the markets for them within and outside the country are limited. On the other hand, the heavy immigration of the last ten years has created an expanding market for table grapes.

Strangely enough, the cultivation of the fig and olive,

Palestine's traditional fruit trees, has shown no marked progress. Nor are the varieties grown in Palestine of the best. The figs are usually eaten fresh from the tree, and we found them large and delicious, but they are much less attractive when dried. Sporadic efforts to introduce Smyrna figs and better varieties of Greek olives have not been particularly successful. The reason may be that modern Jewish agriculture has not been attracted by these fruit cultures. The fig and olive have remained essentially "Arab trees," and the Arab peasant usually prefers to do as his forefathers did rather than introduce new methods. Nevertheless, the area under olive trees is considerable and on the increase. The olives are used not only as food but also—and largely—for the production of oil and soap. Palestine's traditional soap industry is in Nablus, the Biblical Sichem. Though its products can be found in all Eastern markets, the soap industry of Nablus has declined since the war of 1914-18. The more modern soap factories established by the Jews are using copra and palm oil from India and sunflower oil from Russia, along with local olive oil.

Most of Palestine is not suited to deciduous fruits. Jewish settlements in northern Palestine where the climate is somewhat cooler are, however, successfully cultivating apples and pears of exceptionally large size. The peach and apricot, which are the standard fruits of southern Syria, are also cultivated in the Galilean villages.

The Jordan Valley produces sizable quantities of bananas which are all marketed in Palestine. Jewish growers have introduced persimmons, avocadoes, cheremoyas and papayas, which seem to prosper in the Maritime Plain and the Jordan Valley.

[98]

5

5
5
NEW FARMERS IN A NEGLECTED LAND

While Arab agriculture in Palestine is concentrated to a great extent on the cultivation of cereals, Jewish farmers stress the intensive branches of agriculture based at least partly on irrigation. The larger collective farms grow cereals on part of their land but prefer diversified farming as offering better results, more security and opportunities for a larger population. Even where irrigation is not available, they give much attention to fodder plants as a basis for intensive dairy and poultry farming.

The area of Palestine planted with wheat slightly exceeds that planted with barley. In 1938 these two crops were planted on 4,116,000 dunams, or 1,029,000 acres, which is about 60 per cent of the total cultivated area of Palestine. An African grain sorghum, mainly used for poultry although it is still used in the poorer sections of Palestine for humans, was planted on 248,778 acres in 1938.[3]

The general development of Jewish agriculture for the last two generations is given in the accompanying table. (A dunam is one-fourth of an acre.)

Year	No. of Agricultural Settlements	Inhabitants	Land Area (in dunams)
1899	22	5,000	300,000
1914	44	12,000	400,000
1930	107	45,000	1,050,000
1936	203	98,300	1,480,000
1939	252	137,000	1,650,000

In spite of this remarkable expansion, the country is still far from self-sustaining with respect to food. Some of its foods —cereals and flour, eggs and meat, and even certain kinds of

[3] Palestine, Office of Statistics, Statistical Abstract, 1940.

555
5
5
5
5
5
5
5
5
5
5
5
5
5
5
5
5
5
5
5

vegetables—were imported in considerable quantities before the war. But even countries like the United States import certain agricultural products, and agricultural self-sufficiency can hardly be expected of so small a country as Palestine. Great efforts have been made during the war to increase the production of foodstuffs and thereby save shipping space for military purposes. As a result the production of cereals and dairy products has risen about 30-35 per cent above the pre-war level while the output of vegetables, which are playing an ever more important part in the food balance of the country, has been more than doubled. This is a further indication that the limits of agricultural expansion are far beyond the achievements of the past.

Fishing

The Jewish colonizing agencies are much concerned with the development of fishing as an aid in improving the food balance of Palestine. A number of fishing villages have been established on the shores of the Mediterranean and the Sea of Galilee. They are organized on a semi-agricultural plan which protects them against depression in poor fishing seasons and provides supplementary work for their members during slack periods.

Before Jewish settlements took up fishing, the Arab fishermen of Palestine used primitive boats and inefficient methods. Now some of them have been encouraged to use the same equipment and methods as their Jewish neighbors. A number of Jewish villages have also stocked ponds with fish. At Aqaba in southern Trans-Jordan we were told of, and saw, swarms of fish in the waters of the Red Sea. Since modern methods of refrigeration are not yet fully developed in Palestine,

Jewish leaders hope to establish a sizable fishing industry near Aqaba and transport the catch to Palestinian cities which are only a few hours away by improved roads. This plan should become feasible after the war.

Reforestation

The restoration of Palestine's rocky hills will require a great deal of reforestation. If carried out on a large scale, reforestation will increase the land's absorption of rainfall and help to improve the climate. In addition, it will eventually supply the country with considerable quantities of lumber. Some Jewish settlements are already using eucalyptus trees for fuel and lumber. In addition, it is beginning to be realized that forests may serve indirectly as a source of food. In Cyprus, extensive groves of carob trees provide an important source of concentrated fodder. The pods are sweetish and can also be made into a form of sugar. Some Palestinian collective farms buy these pods from Cyprus as food for their milk cows. Carob trees can thrive with little care on the rocky slopes of Palestine, and there is now much more interest in planting them.

According to the latest official statistics there are 741,504 dunams or 185,108 acres of land designed by the Palestine government as "Forest Reserves." In addition, there are 39,887 dunams or 9072 acres of forests belonging to private persons and organizations, most of the latter having been planted by the Jewish National Fund.

There is a considerable difference between a "forest area," as the term is used by the Palestine government, and a "forest" in the usual sense of the word. A "forest area" is a stretch of land closed, at least theoretically, to goats and fuel gatherers in order that an opportunity may thus be given for the growth of native vegetation and trees. On certain of these areas the government has planted new trees, some of which have now reached a considerable size. The forests of the Jewish National Fund are planted more densely and more nearly approach our conception of woods. While there may be various opinions regarding these two systems, there is no doubt that both government and Jewish organizations are doing excellent work in this important field.

We have attempted a brief and necessarily incomplete review of agricultural development in Palestine. But a much longer and fuller account would hardly convey an adequate impression of the transformation now in progress. Steadily and with increased speed the country is emerging from a backward, low-yield agricultural economy, dependent chiefly on grains and olives, and is evolving toward a modern, scientifically directed and richly diversified economy with fruits, vegetables, poultry and dairy products playing an ever greater role. The wooden plow is yielding to the tractor, the flail to the threshing machine. Rural Palestine is becoming less and less like Trans-Jordan, Syria and Iraq, and more like Denmark, Holland and parts of the United States. Continued Jewish initiative and a progressive policy on the part of the government will further accelerate this remarkable evolution.

CHAPTER VII

THE INDUSTRIAL DEVELOPMENT OF PALESTINE

. . . .

BEFORE the modern Jewish resettlement of Palestine, the country had no industry worthy of the name. There were a number of ancient soap factories in Nablus and a few small flour mills in Jaffa and Haifa. Several hundred Arab and Jewish artisans produced beads, mother-of-pearl ornaments and olivewood carvings which appealed to the religious sentiments of Christians and Jews. These were sold to tourists or sent abroad and together with olive-oil soap were the country's only industrial exports.

The task of developing modern industries in Palestine was no less formidable than that of introducing advanced agriculture on its old wasted lands. A very limited domestic market, abnormal credit conditions, lack of tariff protection and many other serious handicaps had to be overcome. Nor was there any such century-old development of arts and crafts as had preceded and stimulated the growth of modern industry in England, France and Germany. Lacking this technical tradition, such new lands as Australia and Argentina have received government support for their industries—but such help has been almost completely lacking in Palestine.

The pioneer industrialists were confronted by a number of serious problems. While many Eastern European Jews were versed in various handicrafts, skilled workers were rather exceptional among the early immigrants to Palestine. But the savage persecution of the last decade served to uproot even those European Jews who seemed to be deeply entrenched in the economic life of their native lands. Most of the young people who went to Palestine as labor pioneers in the days before 1933 belonged to middle-class families who had lost their economic bearings before the advent of Nazism. They had had no working experience in their early youth, and since their training before departure to Palestine was limited to the rudiments of farm work, they were unacquainted with the intricacies of modern machine-driven industry. Even when workers recruited by the first Palestinian industrialists had had some industrial experience, they had to be retrained in new industries utterly strange to them. This at first made their work less productive than that of European workers in parallel occupations, and put the Palestinian industrialist in an unfavorable position as compared with his competitors abroad.

Other difficulties were the limited quantity and variety of local raw materials and the lack of cheap mechanical power. Even after the generation of electricity from water power and the establishment of power stations using fuel oil imported from Iraq, the costs of electricity remained comparatively high.

Owing to its rapid and recent development, Palestine has not been able to build up a strong credit system based on the accumulated savings of its own population. The savings of Arab fellaheen, if any, are kept in jars and secret hiding places. A Jewish immigrant does not, as a rule, have savings that can be diverted into the general credit system of the country. He must invest whatever surplus he can accumulate in his own house, farm or workshop. Under such conditions, industrial credit must be obtained from foreign sources, and imported capital is generally more expensive than domestic savings. Private credit is hardly available in Palestine at an interest rate of less than 8 per cent to 9 per cent a year, and even under these conditions it was at first hard to obtain. Most industrialists were thus forced to rely wholly on their own resources, and large-scale development was severely hampered.

Palestine's Trade Status

The greatest handicap to Palestine's industry is the country's peculiar status as a mandated territory whose administration is formally responsible to the League of Nations. According to Article 18 of the Mandate there shall "be no discrimination in Palestine against goods originating in or destined for any of the countries which are members of the League of Nations." This provision was extended to the United States by the American-British Palestine Mandate Convention of December 3, 1924. Woodrow Wilson and the other founders of the League had optimistically expected all the leading countries of the world to adopt the principles of free trade and thus naturally applied them to the mandated territories which had been placed under the supervision of the League. Certainly, such a general policy is admirable in theory. The democratic nations now realize to their sorrow that if the liberal policy urged by Wilson at the end of the last war had actually become the leading principle of practical world politics, the present holocaust might have been prevented. But it was neither just nor practical to impose free trade upon a small country at the time when all the larger countries of the world protected their industries by tariff walls and import quotas. Unable to bargain with or retaliate against a country discriminating against it, Palestine was forced to accept considerable quantities of foreign goods without any opportunity to induce others to admit its own products in exchange. This was not only one of the greatest difficulties in the way of Palestine's industrial development, but also the chief reason for the depression in the citrus industry before the war.

Had Great Britain been actively interested in the industrialization of Palestine, she probably could have gained the consent of the League to certain amendments in Palestine's trade status. A number of legal experts have even asserted that Article 18 can be interpreted in a fashion which would allow Palestine to grant a general tariff reduction to countries accepting a definite percentage of Palestinian goods. An arrangement of this sort, since it is not directed against any country in particular, could not be considered willful discrimination. In fact, the country that would gain most from such a graduated tariff would be Great Britain, to which over 60 per cent of Palestine's total exports were sent in pre-war years.

Article 18 was only one of the causes of Palestine's adverse trade balance. Before 1914 a marked excess of imports over exports was quite natural, in an undeveloped country largely dependent upon tourist traffic and upon remittances sent from abroad to support institutions of a religious character. After the First World War a new factor made Palestine still more dependent on imports: Large-scale reconstruction work was begun and machinery had to be brought from abroad. For example, orange groves must be cultivated for six or seven years before their fruit can be exported, and during this preparatory period large amounts of imported equipment are used—well-boring machinery, tractors, irrigation pipes, fertilizers, lumber, trucks, etc. Thus even citrus plantations contribute to the country's adverse trade balance, as do industrial enterprises until they are completely established. Indeed, in the period between 1929 and 1940 Palestine's imported industrial machinery was valued at £P5,900,000, or $29,500,-000. A still larger sum was spent for transport vehicles and agricultural machinery.

Growth of Industries

The following figures on Palestine's foreign trade during the eight years before the war show that the peak was reached in 1935, the year in which the largest number of Jewish immigrants entered the country.

PALESTINE'S FOREIGN TRADE 1932-39 (in £P)*

(Statistical Abstract of Palestine, 1940, pp. 4 and 61)

Year	Imports (£P)	Exports (£P)
1932	7,768,920	2,381,491
1933	11,123,489	2,591,617
1934	15,152,781	3,217,562
1935	17,853,493	4,215,586
1936	13,969,023	3,625,233
1937	15,903,666	5,813,536
1938	11,356,963	5,020,368
1939	14,632,822	5,117,769

* Palestinian pounds approximately equal to English sterling.

Trade figures for the war years of 1940-41 show a great decline, particularly in exports. Citrus had accounted for

65 per cent of the pre-war export and only negligible quantities of it have since been sent out of the country. According to official statistics, total imports to Palestine in 1941 amounted to £P11,035,454, as compared with £P12,038,028 in 1940; the corresponding figure for exports is £P1,361,895 and £P2,119,584. The current exchange value of the Palestine pound is $4.02.

The figures for the years 1940-41, however, omit many important items, particularly in exports. Palestine's industrial contribution to the war effort is hardly reflected in them. Military orders are delivered to the purchasing agents of the military authorities in Palestine, and when exported are not registered in customs. The export of minerals extracted from the Dead Sea and of precious stones processed by Palestine's growing diamond industry are, likewise, not included in the above figures.

The rapid growth of Jewish industry in Palestine is shown by the accompanying table. (Statistical Abstract of Palestine, 1940.)

	1921	1930	1933	1937	1942
Number of enterprises	1,749	2,475	3,388	5,606	6,500
Workers employed	4,434	10,968	19,595	30,040	53,000
Capital invested (£P)	485,000	2,234,000	5,371,000	11,637,000	26,000,000

This table includes many small shops employing one to five workers; but it does not include the building trades, which normally absorb a very considerable amount of Jewish capital and labor.

To give a clearer picture of Palestine's industrial development, let us divide the enterprises included in the 1937 figures into two categories—those employing less than ten people and those employing at least ten workers. Only shops belonging to the second category can be designated as factories in the usual sense of the word.

PALESTINE'S JEWISH INDUSTRY IN 1937

	Number of Enterprises	Number of Employees	Capital Invested £ P
Smaller shops (under 10 workers)	4,050	8,022	573,330
Factories (over 10 workers)	1,556	21,964	11,063,791

No industrial census has been taken in the last few years. It is conservatively estimated that at the beginning of 1942 the number of factories employing over ten workers was 1800; the number of their employees was 45,000; and the capital invested in them was £P14,000,000 or $70,000,000.

To Americans these figures may seem insignificant. Actually, when contrasted with the backward state of the rest of the Middle East, they are indicative of exceptionally rapid industrial development. According to the latest statistics, 89 per cent of the capital invested in Palestinian industry and 77 per cent of its labor is Jewish; Palestinian Arabs have been stimulated by this Jewish initiative and are estimated to have established about 2000 industrial plants, generally much smaller than the average Jewish enterprises.

Branches of Industry

A number of industries have grown up around Jewish agricultural development in Palestine. The first step in this direc-

[109]

tion was taken when Baron Edmond de Rothschild built wine cellars in Rishon-Le-Zion. Soon afterwards the Baron tried to create a perfume industry based on the cultivation of geraniums, jasmin and other flowers. He also established a glass factory to produce the bottles required in his extensive wine cellars.

A considerable part of Palestine's industry today is directly related to agriculture. There are many plants which process farm products, and others which supply farmers' needs. Flour mills, bakeries and biscuit factories are examples of the first type, as are vegetable oil factories which utilize Palestinian olives, sesame and soya beans along with imported oil seeds.[1] The canning of grapes, citrus fruit and vegetables is becoming an increasingly important industry.

Palestinian plants manufacture various types of agricultural machinery, pumps and irrigation pipes. Before the war plywood imported from Russia and Rumania was used to make the crates in which Palestinian citrus exports were packed.

Building and Building Industries

Four hundred thousand Jewish immigrants entered Palestine during the last twenty years, and as a natural result building became one of the country's chief industries. During the peak year of 1936 building projects gave employment to 19,000 Jewish workers—over 13 per cent of the Jewish laborers then in the country. About 6 or 7 per cent were occupied in the production of building materials.

[1] The Shemen oil factory in Haifa, one of the oldest and largest industrial undertakings in the country, exports considerable quantities of oil, soap and cosmetics to the neighboring countries.

THE INDUSTRIAL DEVELOPMENT OF PALESTINE

When the Jewish settlers came, they imitated the Arabs, building crude stone houses by manual labor. Those few Jews who had building experience in Europe were utterly ignorant of the specific problems confronting builders in Palestine. But Jewish workers were intelligent and eager to learn. Within a comparatively short time after the last war a large number of skilled building workers had been trained. At the same time architects and civil engineers educated in technical institutions abroad became more versed in the use of Palestinian materials, and worked out methods of adapting Western techniques to the local climate and traditions. Palestine now possesses an excellent modern school of architecture, as exemplified by the residential sections of Haifa and Tel Aviv and such monumental structures in Jerusalem as the Rockefeller Museum, the offices of the Jewish Agency, the Rothschild-Hadassah-University Hospital and the American-sponsored YMCA building. Almost all the building materials now used are of local origin. Jerusalem and Haifa, situated in hilly districts and surrounded by quarries, utilize limestone, while the towns and villages of the Maritime Plain prefer bricks and cement blocks. All Jewish buildings are equipped with modern plumbing and electric fixtures, and lately well-to-do Arabs have begun to imitate their Jewish neighbors in installing modern conveniences.

Large numbers of Jewish workers are employed in building roads, streets and other public improvements. The highways connecting the most important cities of Palestine were built at government expense, but secondary roads to Jewish villages have frequently been built wholly or partly at the settlers' expense.

Many of Palestine's Jewish building workers are organized

in co-operatives which undertake jobs on a collective basis. By rationalizing and intensifying their work, such collective groups are able to attain substantially higher wages for their members than they would receive if they worked on an individual basis. The Histadrut, or General Federation of Jewish Labor, has a contracting office which undertakes large-scale public projects throughout the country, employing architects and civil engineers in addition to laborers. The profits, if any, belong to the Labor Federation. When the Arab riots of 1936-39 and the resulting economic depression cut down private housing, the Histadrut contracting office carried out a number of government projects, most of which were connected with public security. These included the building of strategic highways, police barracks and the "Teggart Wall" on the northern frontier of Palestine. Since the outbreak of the present war Jewish workers have played the chief role in the construction of fortifications throughout Palestine.

As a natural concomitant of the building industry several large and many small brick and cement factories have sprung up, as well as others for the output of lime, gypsum, alabaster and other materials. The plant of the Nesher Cement Company near Haifa produces 400,000 tons annually and ranks among the largest and most modern factories of its kind.

An interesting feature of Palestine's Jewish building industry is the comparatively important place taken in it by women. Women have become experts at certain specialties such as tile-laying. One often sees women building brick walls alongside male comrades, just as they take their full share in agriculture, research and national defense.

THE INDUSTRIAL DEVELOPMENT OF PALESTINE

Palestine Industry in the War

The rapid development of Jewish industry in Palestine has proved extremely valuable to the cause of the United Nations. The foundations laid during the last twenty years have made possible the establishment during the war of more than three hundred important new industrial undertakings, including chemical and metal works, textile factories, pharmaceutical laboratories, electrical shops. Palestinian Jewish industry now employs 45,000 people, and its output has gone up by 75 per cent since the outbreak of war. Army contracts totalled £P1,000,000 in 1940, £P4,000,000 in 1941, £P10,-000,000 in 1942.

Goods produced for the army range from tin water-bottles and clothing to machine and weapon parts, and from potatoes to marmalade. Through the United Kingdom Commercial Corporation important supplies go to Turkey, including boots for the army (affectionately called "Churchills" by the Turks), canvas for tents and heavy harbor equipment. The largest foundry and the only glass factory in the Middle East are in Palestine—both co-operative enterprises. Scientific institutions and personnel increasingly turn to the war effort. Important work of a specialized character is done in the laboratories of the Hebrew University, in the Daniel Sieff Research Institute (Dr. Chaim Weizmann is its director) and in the laboratories of Professor Goldberg, formerly of Zeiss, at Tel Aviv.

Jewish technical and skilled personnel is greatly in demand throughout the Middle East: A wide network of roads in Syria was built by Jewish contractors, engineers and foremen (the unskilled labor being local); a bridge across the

[113]

Euphrates was designed and constructed by Palestinian Jews.
The great refineries of the Anglo-Persian Oil Company carried through extensions made urgently necessary by the
German occupation of Rostov, with the help of hundreds of
technicians and artisans recruited in Palestine. Jewish experts and skilled workers are to be found in Egypt, in Eritrea,
in Abyssinia; a few have even been called to India and
Ceylon.

Co-operatives in Industry

Zionist public funds subsidize mixed farming during its
initial stages, but industrial enterprises have grown up with
very little help. About 90 per cent of them were founded with
private money, and most of them are conducted on conven-
tional capitalistic lines.

The spirit of social adventure expressed in the far-reaching
innovations in the agricultural villages has, however, not
failed to influence Palestine's industrial development. Despite
the preponderance of privately owned factories, there are an
unusually large number of industrial co-operatives which
have met with more success than similar enterprises in other
countries. They include 59 carpentry and 36 metal shops, 38
food processing businesses (bakeries, canning and preserving
factories), printing shops and a number of establishments
producing cloth, clothing and shoes.

Co-operatives practically monopolize local transportation
by buses and trucks. Some of them have increased their capital many times over and their membership enjoy a comparatively high standard of living.

The success of Palestinian industrial co-operatives seems
to be due to the devotion of their membership and to the in-

fluence of the Labor Federation, which does everything possible to keep co-operatives from degenerating into limited partnerships.

The Semi-industrial Village

In my opinion, Palestine's most important social achievement in the industrial field is the introduction of industries into the co-operative settlements which were originally intended to be purely agricultural. Collective farms have found it to their economic advantage to establish carpentry or printing shops, trucking services, shoe repair stores or small factories. Having these added sources of employment and income, the farms become less dependent on the vicissitudes of the weather. The industrial plants provide work during the slack season between planting and harvest. When work in the fields is at its peak, industrial activities are reduced to a minimum and only skeleton staffs of skilled workers remain in the plants. Palestine is particularly well suited to the development of a rural-industrial economy. The smallness of the country makes transportation simple, and electric power is available in villages as well as towns.

According to the latest statistics, the co-operative villages now derive about 22 per cent of their income from industrial undertakings, and there is every reason to expect increased industrial development in the future. The idea of introducing a certain amount of industrial work into agricultural communities is not new, it has been advocated by persons as much at variance with each other as Prince Kropotkin, the anarchist, and Henry Ford, who, in fact, supports several remarkable experiments along these lines. We are indebted to the Palestinian pioneers for further practical application of this forward-looking idea.

Another new industrial form found in Palestine is joint ownership of industrial enterprises by labor groups. In such partnerships the capitalist usually receives limited interest on his investment, and works in the factory as an industrial or commercial manager side by side with his labor partners who do the physical work. The initial success of this type of arrangement augurs well for its future.

Electrification of Palestine

The most important single factor in the development of modern industry in Palestine was the electrification of the country by the Palestine Electric Corporation. The Corporation was founded by the late Pinchas Rutenberg, an able engineer and strong-willed personality who made the electrification of Palestine his life work. Originally supported by Zionist funds, Rutenberg was later able to interest private capital in his project. The share capital and reserves of the Corporation now amount to £P4,000,000 or about $16,000,-000 at current rates of exchange. Dividends have been paid regularly for the last ten years. The Rutenberg concession applies to the whole of the country with the exception of the Jerusalem district, where a smaller company operates on a grant given it by the Turkish government before the First World War. The Palestine Electric Corporation supplies 92 per cent of the total electricity used in Palestine. The current is derived from a hydroelectric station on the Jordan and from power stations operated by fuel oil driven from Iraq through the pipe line to Haifa.

The consumption of electric power has increased rapidly

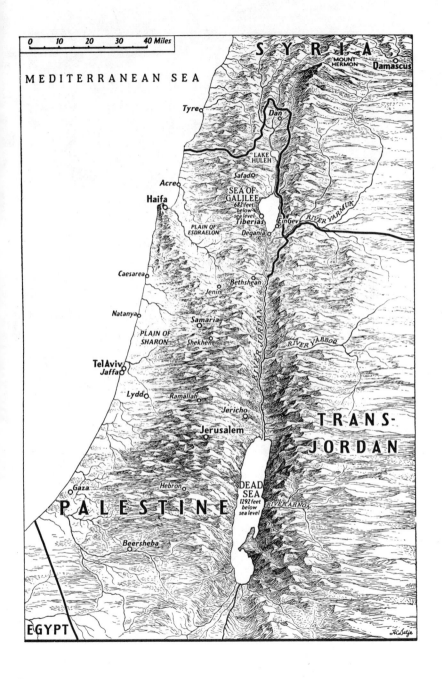

0   10   20   30   40 Miles

MEDITERRANEAN SEA

SYRIA

MOUNT HERMON

Damascus

Tyre

Dan

LAKE HULEH

Safad

Acre

SEA OF GALILEE
682 feet
below
sea level

Haifa

Tiberias

EinGev

RIVER YARMUK

PLAIN OF ESDRAELON

Degania

Caesarea

Bethshean

Jenin

Natanya

Samaria

RIVER YABBOK

PLAIN OF SHARON

Shekhem

RIVER JORDAN

Tel Aviv

Jaffa

Lydd

Ramallah

Jericho

TRANS-
JORDAN

Jerusalem

Gaza

Hebron

DEAD
SEA
1292 feet
below
sea level

RIVER ARNON

PALESTINE

Beersheba

EGYPT

1a—The sand dunes on the Mediterranean shore where Tel Aviv was built in 1909.

1b—Thirty years later, the heart of the modern all-Jewish city of Tel Aviv (an aerial view).

2a—Bare, rocky hills outside Jerusalem, where 20 years ago the middle class residential suburb of Beth Hakerem was founded.

2b—The same spot, today, transformed by evergreens, gardens and modern dwellings.

3a—Typical Palestine tragedy in land use—soils washed from hills to bed rock.

3b—Section of the Balfour Forest—soil restoration by afforestation in the last two decades.

ta—Beth Yoseph in the Jordan Valley—a settlement erected within a stockade during the 1937 riots. Note the water tower, typical of Palestine colonies.

4b—Beth Yoseph, two years later, expanded far beyond its original walls.

5—Nahalal, a large co-operative village founded in 1921 on malarial swamp in the valley of Esdraelon near Nazareth.

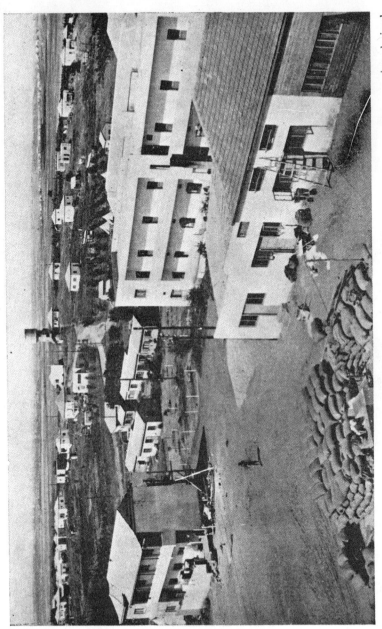

6—Kfar Hogla, one of many villages dotting Emek Hepher along the Mediterranean coast, a swampy region drained only ten years ago.

7—Once a successful lawyer in Czechoslovakia, now a farmer in the Palestinian village of Nira.

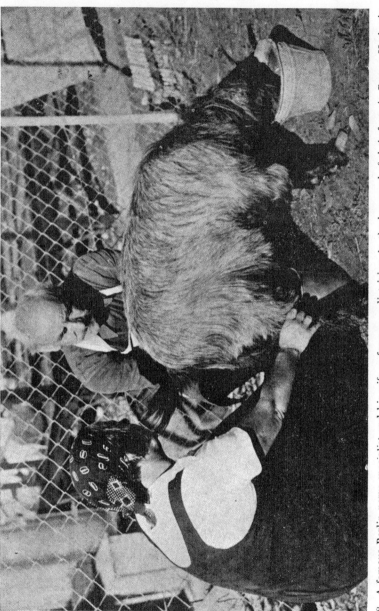

8—A famous Berlin nerve specialist and his wife, a former pediatrician, in the barnyard of their farm in Ramat Hashavim.

9—Surplus citrus fruit being turned into fodder for cattle.

10—Lodzia, one of the modern textile factories near Tel Aviv, named after Lodz, Poland, once a Jewish textile center.

11—Women take men's places behind the plough to relieve the manpower shortage resulting from large scale enlistments.

during recent years. The accompanying table shows the total amount of electric current sold by the company since 1932 and the specific amounts used for irrigation, industrial power, lighting and other purposes (mainly electrical appliances in homes):

SALES OF ELECTRIC CURRENT BY PEC (in kw hrs.)

	Total	Irrigation	Industry	Lighting, Etc.
1932	11,590,350	4,399,533	4,058,629	3,132,188
1933	20,136,839	9,029,436	6,575,526	4,531,877
1934	34,385,515	14,281,892	9,855,466	10,248,157
1935	50,362,193	16,134,366	17,166,602	17,061,225
1936	65,495,957	24,122,151	18,710,245	22,663,561
1937	71,265,889	25,334,888	20,314,114	25,616,887
1938	72,253,610	25,502,139	20,180,322	26,571,149
1939	84,077,141	28,504,454	25,104,437	30,468,250
1940	93,873,482	28,234,021	35,264,247	30,375,214
1941	103,031,247	33,297,560	32,865,401	36,868,286

The Dead Sea Concession

Like the Palestine Electric Corporation, Palestine Potash, Ltd., the holder of the Dead Sea Concession, was established largely as the result of the efforts of a single individual. Mr. Moses Novomeysky, formerly a mining engineer in Siberia, worked tirelessly for many years on investigating the Dead Sea's resources and obtaining the concession to exploit them. We were noisily reminded of his life work each morning about 4:00 A.M. as the great convoy of trucks loaded with valuable Dead Sea minerals roared up the hill past our temporary home in Jerusalem on their way to the ships which would carry their contents abroad.

The Dead Sea, it has become clear, contains a practically

[117]

inexhaustible quantity of valuable minerals. Evaporation is intense in that hot valley 1300 feet below sea level and the Potash Syndicate has constructed huge evaporating pans extending over thousands of acres around the flat shores of the Dead Sea. In these pans the brine is concentrated by sun and wind, and is then dug out by machines. The company employs both Arab and Jewish workers. The acclimatization of Jewish workers from Central and Eastern Europe to the great heat of the Dead Sea region presented a serious problem, similar to that experienced during the building of Boulder Dam in a scorching desert canyon. The problem has been most successfully solved at the Dead Sea, for no epidemics have occurred among the workers, and the mortality rate is not much higher among them than among those employed in cooler parts of Palestine.

Since the outbreak of the war Palestine Potash has greatly increased its production.[2] The Dead Sea is the largest source of potash and bromine in the British Empire. It contains many other deposits which are yet to be exploited: common salt, immense quantities of magnesium and other minerals which can be most important in the development of a large-scale chemical industry.

[2] EXPORT OF THE PRINCIPAL DEAD SEA PRODUCTS (in tons)

	Bromine	Potash
1935	403	18,124
1936	478	19,793
1937	533	29,110
1938	481	47,496
1939	589	63,527

THE INDUSTRIAL DEVELOPMENT OF PALESTINE

The Harbors of Palestine

Up to the last ten or fifteen years Palestine's harbors were extremely primitive. Most foreign trade went through Jaffa, where incoming ships must anchor about a mile from shore and passengers can be landed or goods loaded only by the use of small lighters. Palestinian exporters, particularly of citrus, constantly complained that these inefficient conditions were detrimental to their business.

A similar situation prevailed in Haifa till 1933 when the large modern harbor was completed. Even this, however, could not meet the growing needs of the country. Most citrus exports went through the less modern but more convenient harbors of Jaffa and Tel Aviv, which are nearer the citrus groves of Judaea and the Sharon Valley.

The harbor of Jaffa has been used since ancient times, but the Tel Aviv port is of very recent origin. Three years before the outbreak of the war, the citizens of Tel Aviv began to build their own harbor at the estuary of the Yarkon River in the northern part of the city. This undertaking was stimulated by Arab terrorist activities, which made it unsafe for Jewish exporters and port workers to enter Jaffa. The Tel Aviv port was financed by the voluntary contributions of the Palestinian Jewish community.

Strenuous efforts have been and are being made to develop a Jewish merchant marine and train Jewish harbor workers. Jews have for centuries had very little connection with the sea, and special schools have therefore been opened in Palestine to train Jewish boys as fishermen, sailors, navigators and ship mechanics. Many hundreds of these young people are

[119]

now serving in the Royal Navy and United Nations merchant marine.

Before the war, Palestine, because of its geographic position, became an important center of world air communications. Extensive airfields, hangars and repair shops were built to accommodate the growing air traffic from Europe to the East and Australia. Air traffic will undoubtedly increase greatly after the restoration of peace, and Palestine should reap considerable benefits in employment and trade. As we watched the giant planes sweep out of the skies and with a great noise and splash land on the placid Sea of Galilee, we thought of the remarkable prophecy of Ezekiel many centuries ago:

"I heard the noise of their wings, like the noise of great waters."—Ezekiel 1.24

The instruction of young Jews in mechanical trades is considered one of the most important tasks of the Jewish Agency and other organizations interested in the upbuilding of Palestine. Their efforts in this direction have attained considerable success, and a large number of very efficient chauffeurs, bus drivers, repair men, flyers and air mechanics have been trained. Thousands of Palestinians now serve with the British army in all sorts of auxiliary capacities. The experience they are thus acquiring will prove a distinct asset in the great reconstruction tasks of post-war days.

CHAPTER VIII

THE BUILDERS OF THE NATIONAL HOME

. . . .

IN THE course of studying Palestinian agricultural settlements, I became interested in the character and background of the people who founded them and are responsible for their success. Most of us have been wont to think of the Jews as belonging to the market places of the world, and it was startling to find them upon the land, engaged in a co-ordinated colonization effort that, in my opinion, is the most successful of modern times.

The transformation of predominantly urban European Jews into successful Palestinian farmers is an astonishing feat. In order to understand it, we must remember that Jews were not an urban people during their first thousand years in Palestine. The overwhelming majority of them were peasants. They did not excel as tradesmen, and Phoenicians and Philistines, Idumaeans and Arabs controlled most of Palestine's external trade during the various periods of its history. Jews became a city people only after their exile from Palestine. They were forced into trade because most of the countries which gave them refuge would not tolerate foreign agriculturists on their soil. There was no place for outsiders in feudal agriculture, but feudal society tolerated and sometimes even welcomed foreign traders who enriched the treasuries of rulers. Simply in order to exist the average Jew had to become a competent trader. This was definitely the result of historic adaptation rather than of an inherent national characteristic. It is interesting to note that in the Near East today Jews are not considered unusually able merchants. A Turkish proverb says that "One Armenian trader is as good as two Greeks, and one Greek as good as two Jews."

Accomplishments such as I have seen in Palestine can be attained only when the human spirit is fired by an ideal which reaches beyond the individual to the group and on into the future. Since the destruction of the Temple nearly two thousand years ago, the longing for Palestine has been ever present with the Jew. It was this undying hope to return to the land of his fathers that buoyed him up during centuries of persecution. The devout have ever prayed for the restoration of the Jewish people to Palestine, but it was only about sixty or seventy years ago that Jews began to do practical work in Palestine in order to make the return possible. Zionism has offered persecuted Jews the opportunity to make their ancient dream come true, just as other great dreams of the world have been realized when men worked for them with devotion, intelligence and self-sacrifice.

Training of Pioneers

The first Jewish immigrants were far from adequately prepared for the arduous work of reclaiming Palestine. Idealistic young people from middle-class families, they came directly from their homes, shops or universities and had to become adjusted both to a new land and to new and difficult work. Many failed to make the adjustment; some ultimately left for other countries which seemed to offer greater and easier opportunities; others became planters of the usual colonial type, relying on the cheap labor of "natives." A comparatively small number of the early colonists succeeded in establishing themselves on Palestine's soil and living by the fruits of their own labor.

After the Balfour Declaration of November 2, 1917, the training of pioneers was organized in a more rational way. The Zionist movement entrusted this task to a special organization called "Hechalutz" ("The Pioneer"). In it all the young people were enrolled who for idealistic or economic reasons, or both, were intent on going to Palestine. In addition to training them in agriculture, Hechalutz surpervised their psychological preparation for the tasks awaiting them in Palestine. The first step in their training was to learn Hebrew, the traditional tongue of the Jewish people and the common language of the Palestinian Jewish community. They received instruction in the geography and nature of Palestine, and were energetically educated in the spirit of discipline and co-operative effort. After these preliminary steps, those who were found fit to go to Palestine were sent to large training farms. Under the instruction of local agriculturists and special instructors from Palestine, they underwent a discipline of hard work and simple, strenuous living. Some of them dropped out in the course of this rigorous training, but those who went through it to the end turned out to be exceptional human material.

The "Hechalutz" organization also maintained a number of shops where the prospective pioneers were taught trades considered essential to the development of their new country. Of course, not all the immigrants received this special preliminary training. Some were "capitalists," which, in the terminology of Palestinian immigration laws, meant people possessing a minimum of £P1000 ($5,000) and therefore presumably able to establish themselves with their own means in farming, trade or industry. There were also qualified artisans who, because of their special skills, were admitted without the usual preparation prescribed for the unskilled pioneers. A considerable number of professionals (physicians, lawyers, dentists, rabbis, students, etc.) were likewise not required to undergo special training. Nevertheless, about half the Jewish immigrants who came to Palestine after the issuance of the Balfour Declaration and before mass refugee immigration started in 1933 had gone through preparatory training in Hechalutz groups abroad. Owing to this organized training, the human element in the upbuilding of Palestine fully matches in its constructive capacity the best of the pioneers who developed the New World.

Means and Methods of Settlement

As a general rule, the young pioneers had no means of their own, and sometimes even their passage had to be paid by the Jewish Agency. They could not be settled on the soil unless the costs were borne by public funds. Fortunately such funds had been established while methods of co-operative settlement had been worked out and successfully experimented with in Palestine before the first World War.

It was at the beginning of this century that the newly formed Zionist Organization began to interest itself in practical work in Palestine. The older settlers, financed partly by their own funds and partly by philanthropists and small pro-Zion groups in Europe, had in most cases gradually become plantation owners, leaving most of the physical work to hired native labor. They were reluctant to employ Jewish workers, and thus could not be counted on to provide economic opportunities for new Jewish immigrants. Some other type of agricultural economy was clearly necessary if a sound foundation was to be provided for a large and sound Jewish National Home. Continuation of the old system could have resulted only in a lopsided agricultural community in which the upper stratum of Jewish planters would rest upon a broad foundation of Arab labor.

Closely connected with this problem was that of land values. With the continuous development of Jewish settlement, land prices rose rapidly and considerable land speculation followed. Had land been permitted to become the private property of farmers settled with the help of the Zionist movement, those farmers, in spite of their good intentions, would finally have been drawn into the vortex of land speculation. Experience taught Zionist leaders that it was imperative to place certain limitations on land ownership to create an agricultural community based on the labor of the farmers themselves.

These considerations led to the establishment of the Jewish National Fund which buys land with money contributed by Jews throughout the world. Land acquired by the National Fund becomes the perpetual property of the Jewish people and is never resold. It is leased to settlers for periods of from thirty-five to forty-nine years, with the assurance that at the end of that period the lease will be automatically renewed for another long period of time, provided the settlers meet certain elementary requirements. These are the payment of a moderate yearly rent (about 2 per cent of the value of the land), and the cultivation of the land by the settler's own efforts without the continuous use of hired labor. Thus the lessees of National Fund lands have practically all the security offered by private ownership and all the usual inducements to improve their land.

Since the last war the Jewish National Fund has spent some twenty-six million dollars for land and improvements. It has bought over 140,000 acres or about 35 per cent of the total area belonging to Jews in Palestine. Twenty per cent of the Jewish land domain belongs to Pica, a semi-private colonization agency established with a philanthropic purpose by the late Baron Edmond de Rothschild. The remaining 45 per cent is divided among several thousand private owners.

Co-operative Farming

A considerable number of communities on National Fund land are co-operative villages known as *moshavim*.[1] Each family lives in its own house and takes care of its own orchard, vegetable garden, dairy and chicken runs. In some moshavim the grain fields, too, are cultivated individually; in others they are cultivated jointly and each villager receives a share of the crop proportionate to the number of acres he has put under grain. All marketing—selling and buying both—is done co-operatively and no private business is carried on. Mutual assistance is the underlying principle of life in the moshav. If a farmer is ill or incapacitated, his neighbors cultivate his land for him without compensation.

In case a member of the moshav feels that he must leave the village, he is entitled to compensation for any improvements he may have made on his farm. If he finds a successor who is acceptable to the community, he may make arrange

[1] The plural of *moshav*.

ments with him concerning compensation. Generally, however, the community chooses the successor and fixes the amount of compensation by arbitration.

In the *kvutzot* [2] or collective settlements not only agriculture but many other aspects of life are carried on co-operatively. There is a communal dining room where all members eat the food prepared in a common kitchen. We found these meals simple but well planned and nourishing. We were especially interested in the "children's house" where the young generation is raised and educated by trained nurses. We were told that the fact that the children live apart from their parents and are brought up by strangers does not weaken the attachment between children and parents. The friction that often occurs when tired parents have full responsibility for their children is completely eliminated. As we looked at the screened-in cribs where babies were enjoying their daily sun baths, we were delighted by one ten-month-old infant whose wide smiles kept revealing two teeth and who waved his arms and legs furiously, as if trying to express some inner joy. Soon we realized that our guide, his father, was the object of this happy demonstration.

The collective colony is one of the most amazing instances of human co-operation in the world. Unless one has talked and visited with these settlers, it is hard to realize how much they gain by their disregard of personal wealth. They have no fear of want, no envy of another man's possessions. They work very hard but suffer from neither unemployment nor fear of unemployment. A livelihood for every family is assured. In addition to these material advantages, the mem-

[2] This is plural; the singular is kvutza.

bers of a kvutza are not bound by the shackles that grip many whose chief aim is wealth and great possessions.

Thinking men now realize that a new design for living must be worked out which will give security to small farmers and laborers. In the period of reconstruction which must follow allied victory, wealth will have to become the servant, not the master of human beings. The communal villages in Palestine are worthy of study as outstanding examples of a type of life in which personal wealth is not the goal, and in which all work for the common welfare and each enjoys the fruits of the community's labor.

Types of Co-operative Settlement

I saw other co-operative groups, some members of which cultivate the soil while others, usually a majority, work for pay in neighboring factories or on private farms. They pool all their earnings in a common chest which supports all members on an equal basis. In prosperous years part of this common revenue is used to enlarge the co-operative farm, erect buildings, equip workshops and, in general, put the community on a stronger economic basis. The progress achieved by such a community enables it to accept new members, some of whom are newly arrived refugees.

From the Zionist point of view this is the most important advantage of co-operative settlements over private farms. When a private farm requires more working hands, its owner is naturally reluctant to admit strangers, unless they are transient laborers with a depressed standard of living and with no share in the further development of his farm. A farming co-operative, however, is not deterred by such considerations, and its working membership grows as rapidly

[128]

as its economic opportunities. For a movement primarily interested in further settlement, this consideration is of paramount importance.

To Americans raised for generations on the principle of "rugged individualism" the form of these settlements must seem very strange—the product of a doctrinaire social philosophy rather than of economic realities. However, anyone who has an opportunity to inspect the meticulous books kept by the collective farms and to interview some of their members soon realizes that theirs is the most practical agricultural setup possible in Palestine. Co-operative settlement has proved cheaper than individual colonization both in initial costs and in maintenance, and has thus naturally become the preferred method for the absorption of new immigrants, the great majority of whom come to Palestine practically penniless.

A very tiny minority leave the collective settlements. This was true even in the period of prosperity between 1933 and 1935 when qualified workers in the cities received unusually high wages and enjoyed a standard of living equal to that of workmen in the same trades in Western Europe. In many cases the members who withdrew from settlements were motivated by a desire to assist parents or other close relatives newly arrived from Europe. Not only do the great majority remain faithful to their settlements, but the second generation has thus far consistently followed in the footsteps of its parents.

Religion of Labor

The co-operative and collective settlements owe their success not only to their economic advantages but also to their members' devotion to their ideas and principles. The pioneers one meets in Palestine have a fanatical belief in their mission as torch-bearers of a Jewish Homeland built on the basis of productive work. Theirs is an ideology which glorifies labor and considers it an essential element in the good life. This belief has enabled the sons and daughters of small shopkeepers to transform themselves into workers willing to do every kind of arduous physical labor required for the upbuilding of Palestine. We were thrilled and inspired by the Jewish young people we found establishing new colonies on rocky hills. Far from considering themselves martyrs, they rejoiced as they toiled at tasks which would have seemed hopeless to less courageous spirits.

This Palestinian "religion of labor" does not differentiate between hired workers in factories and farmers tilling their own fields. Professional people in the cities and artisans are likewise considered workers as long as they live by their own physical and mental effort, without exploiting the labor of others. Thus the working class as understood in Palestine is not limited to hired industrial workers but, at least theoretically, embraces the great majority of the population. The Histadrut, or General Federation of Jewish Labor in Palestine, is not a labor organization in our sense of the word. It includes farmers as well as professional and industrial workers, and gives full membership rights to workers' wives. Its 125,000 members constitute an extraordinarily large proportion of the nearly 600,000 Jews of Palestine. The Histadrut strives to raise the material standards of its members, but this and all its other objectives are constantly subordinated to its chief objective—that of aiding in the establishment of the Jewish National Home.

Social Progress

One is amazed at the great number of well-educated persons who have fled from Europe and become workers in the new Palestine. Many pioneers now engaged in hard work on the soil went to colleges and even practiced intellectual professions in various countries of Europe. In a community of this sort women naturally occupy a place very different from that of their Mohammedan sisters in the Arab lands. We found Jewish women enjoying full equality in Palestine. They possess equal voting rights and may take an active part in all branches of economic life. This would be recognized as an important achievement even in the New World where women from the beginning were on a higher level than in Europe. In the countries of the Near East, whose civilization has from time immemorial been based on the subordination of women, this is a very great accomplishment. The example set by the Jews in Palestine, as well as by Christian women in the Lebanon, is beginning to influence the position of Moslem women in Palestine and in the Lebanon. The long veils are coming off, or at least being draped aside, and Arab women are demanding and achieving more liberty in Palestine and the Lebanon than in Syria or Iraq.

Palestine's Jewish pioneers are intensely eager to prevent their children from descending to the level of the peasant masses of the Near East. Hence they have established a very fine system of elementary and secondary schools, as well as a large network of kindergartens. I found these schools efficient and well staffed; they serve the great majority of Jewish children. Their cost is borne by the Jewish community with a very small subsidy from the government. The young generation which goes through Jewish schools is imbued with a passionate devotion to the land and a strong belief in the ideals of democracy and social justice. In many schools I found children studying half a day and working in the fields the other half. Even small children had their own little gardens and were learning to love and care for the land. I am convinced that these children in time to come will carry on the work of their fathers and even surpass their achievements.

The leaders of the Palestinian Jewish community are proud of the benefits which their work has brought to the downtrodden Arab masses. I was deeply impressed by the fact that on my visits to Jewish settlements at the height of the Arab riots I heard no expressions of ill will toward Arab neighbors who had become the tools of hostile propaganda. The Jews found themselves compelled to defend their life and property against the resurgence of the desert spirit inflamed by skillful propaganda from Nazi and Fascist sources. On the walls of the communal dining rooms we saw photographs of fine-looking youths, all of whom had been killed by terrorists belonging to the bands organized by the ill-famed Grand Mufti. Jewish leaders, however, strongly rejected the principle of "vendetta," or blood vengeance, which is generally accepted in the East. Jews defended themselves courageously against attacks, but they disapproved all attempts to imitate the deeds of the terrorists. They greatly regretted having to divert their attention from peaceful reclamation work to the bloody necessities of defense, and they looked forward eagerly to the time when they would be able to concentrate all their energies on the rehabilitation of the land.

CHAPTER IX

EXAMPLES OF RECLAMATION

. . . .

AN ARAB legend tells us that once upon a time an angel, carrying a sack of stones, flew over Palestine; suddenly the sack burst open and all the stones were strewn about upon the hills. The true story reads very differently. Once upon a time the hills of Palestine were covered with rich red earth and protected by forests, smaller vegetation and terraces. Then the trees were cut down, the terraces were neglected, the fertile soil was washed away by rain and finally only the stones were left on the fields. Some of the earth which washed off the slopes was deposited in valleys, to be cultivated and gradually washed away in gullies; the greater part of the earth was dumped into the Mediterranean during floods. Here the soils were sorted by waves, so that fine-textured particles floated farther out to sea while heavier sands were heaped close to shore and were rolled by the winds and waves into great sand dunes. These in turn dammed up the water channels, so that the coastal plains became deadly marshes where malaria depopulated the land. When Jewish settlers began coming to Palestine during the last few decades of the nineteenth century, the country was inhabited by no more than about 300,000 persons, or less than one-fifth the present population.

[133]

Flying over Palestine early in the summer, we looked down on gray rocky hills over which herds of sheep and goats roamed searching for dry blades and roots. Suddenly, in this drab setting, we saw a green emerald in the landscape below —then another and another. These, we learned, were typical Jewish settlements. Remarkable courage and pioneering effort had gone into creating these green oases. Many of the settlements had had to fight swamps and malaria, while others tried desperately to grow crops by rebuilding terraces on seemingly hopeless slopes. Still others were for many years subject to attacks by nomads and brigands. Some of the colonies, older ones among them, are by now prosperous and comfortable villages, while others are still struggling with the sterile, rocky soil.

Of all the difficulties the first Jewish settlers had to overcome, the worst was malaria. It killed many of their best men, emaciated their children and frequently drove them away from their settlements.

The first to meet this implacable foe, and eventually to conquer it, were the founders of Petach Tikvah, a village established by Jews from Jerusalem in 1878, a few years before pioneers began to come from Europe. The land they bought seemed fertile and well watered, but the Arabs in the neighborhood were weak and unhealthy looking. A doctor whom they consulted told them not to dare to settle in what was obviously a malaria-infested region. Nevertheless they persisted. "And yet," they said, "we will try."

Malaria lurking in stagnant pools along the near-by stream drove them back to Jerusalem after two years of unremitting struggle. But they soon returned, built their houses a few miles away from the river, and proceeded to cultivate their lands fighting malaria in a number of ways. They planted eucalyptus trees in the swamps, for these Australian trees absorb great quantities of moisture. They drained some of the smaller swamps. They ordered and used large quantities of quinine. These measures brought malaria at least partly under control in the Petach Tikvah area. It was eradicated completely after the last war when adequate control measures were instituted as part of the large new Zionist program of reclamation and settlement. These control measures have been carried out during the last twenty years by the combined efforts of many agencies including the American Joint Distribution Committee, Hadassah, the Women's Zionist Organization of America, the Jewish National Fund and the Hebrew University.

Today Petach Tikvah is the largest of the Jewish rural settlements; twenty thousand people live where there were only four hundred fever-ridden fellaheen sixty years ago. The descendants of those fellaheen have remained in the vicinity of Petach Tikvah, prospered and greatly increased in number.

Hadera, founded in 1891, midway between Jaffa and Haifa in the Maritime Plain, suffered from malaria even more grievously than Petach Tikvah. It was built in the midst of a green valley which was in actuality a deadly swamp. In all that area there was no other village! Malaria had made life impossible.

Summer after summer from 1891 to 1906 malaria came to Hadera and carried off men, women and children. Only a few of the original settlers remained, but new people—often the relatives of the dead—came from Europe to take their places. Most of these founders of Hadera were pious Jews possessed of simple faith and dogged courage which grew

[135]

out of their deep religious conviction. Their physician tried to dissuade them from remaining in the village, but they stayed on resolutely. Following the example of the Petach Tikvah pioneers, they planted eucalyptus trees, but the Hadera swamps were so large and deep that the trees could do little to remedy the situation. In 1896 Baron Edmond de Rothschild, the generous benefactor of the early Jewish settlements in Palestine, had some of the swamps drained by laborers from the Sudan. This brought about a certain improvement, but it was not till 1922 that malaria began markedly to subside in Hadera. By 1930, as a result of the introduction of modern malaria control methods in Palestine, malaria disappeared completely in Hadera.

Hadera is one of the healthiest and most flourishing districts in Palestine today. In its cemetery we saw rows and rows of graves in which whole families who had died of malaria in the early years are buried. But leaving the cemetery, we walked down a tree-lined road, past wheat fields and orange groves blooming on what had once been swamp land.

In 1921, thirty years after the founding of Hadera, another village was established among swamps some of which were as deadly as those at Hadera. The co-operative settlement of Nahalal in the Valley of Esdraelon was built near the ruins of two older villages in one of which Germans had lived and in the other Arabs. Most of those Germans and Arabs had died of malaria, the others had moved away.

The pioneers who settled Nahalal knew very well what faced them, but they had had years of experience as farm workers in Palestine and were resolute and sturdy folk, determined to drain the swamps with their own hands and to make the land live again. The experts carrying on the malaria control work of the Zionist colonizing agencies gave them advice and guidance, and as a result they reclaimed their land in a methodical and scientific fashion. They conquered the malaria menace far more quickly than the first generation of pioneers had done. In the summer of 1922, sixty per cent of the settlers were infected with malaria; by the summer of 1924 only a fourth of one per cent were stricken. The disease has now disappeared from the settlement.

We found Nahalal one of the most attractive and interesting of all the villages we visited. In addition to the settlers' families, it housed 220 boys and girls who had been brought out of Central Europe a short time before by the youth immigrant movement of the Jewish Agency. This remarkable child-rescue program known as Youth Aliyah was initiated in 1933 and has by now saved close to 10,000 youngsters of thirteen to seventeen from misery, expulsion and even annihilation. It takes these tormented and hopeless children and, by training them for productive work on the land, turns them into useful, happy, well-adjusted citizens of Jewish Palestine. They are given two years of schooling, spending half a day on general studies and half a day at practical work. Graduates have by now formed new settlements of their own and proved themselves excellent pioneers. It is an American woman, Miss Henrietta Szold, who heads the work of Youth Aliyah, and the American Women's Zionist Organization, Hadassah, which gives most of the funds.

We found the consecration and devotion of these young immigrants at Nahalal most inspiring, and were equally moved by the settlers' eagerness to help them and to welcome increasing numbers of them. We gave a "hitch" in our car

[137]

to one of the young girl students who rode with us for seventy miles to the Rehovoth Agricultural Experimental Station. She was a sixteen-year-old refugee from Germany; her father and brother had been thrown into a concentration camp, her home confiscated by Nazi officers. Her face grew radiant as she told us how much she loved to work on the land and how eager she was to graduate and join her comrades in establishing a new settlement.

Reclamation Work in the Hills

Up to about twenty years ago almost all Jewish settlements were founded in valleys, for it was generally believed that villages in the hill country would have to be aided by subsidies for a very long time. The experience of recent years has, however, proved the feasibility of modern agriculture in the hilly interior which comprises the greater part of Palestine. It is a discouraging and difficult task, this re-establishment of terrace agriculture in the hills, but it has yielded to the courage and the vision of Palestine's youthful pioneers.

One of the first and most important experiments in this field was the collective settlement of Kiriath Anavim near Jerusalem. It was begun in 1920 among desolate, rocky hills on which the only growing things were occasional thornbushes and dwarf trees. Only a few strips of land deep in the valleys were cultivated in a primitive fashion by Arabs. The settlers had to remove stones and debris deposited on their lands by the force of the torrential winter rains. They had to dynamite rocks before they could proceed to build terraces on the slopes. The only fertile soil they found was in the fissures of the rocks, and they had to bring up large quantities of fertilizer before they could plant any trees.

EXAMPLES OF RECLAMATION

Kiriath Anavim now possesses orchards of considerable size planted on the newly built terraces. The products of its vines and its plum, peach and apricot trees are sold in the markets of Jerusalem and Tel Aviv. It has had equal success with its dairy farm, for pasture lasts longer on the cool hillside than in sun-scorched valleys, and the mountain climate has proved favorable to the cows.

Milk production totaled 69,000 quarts in 1923-24, 682,000 quarts in 1936-37 and over a million in each of the last few years. Poultry and honey, vegetables and fodder, are other important products of the settlement. Started by sixty people, it now numbers about three hundred and has operated for the last few years without a deficit. As an experiment showing the agricultural possibilities inherent in the barren hills of Judaea, it must be considered a great success. The Royal Commission which in 1936-37 investigated the causes of the Arab riots visited Kiriath Anavim and found it to be "remarkable testimony to the enthusiastic energy not only of the immigrants, but of those who financed and advised them. Land which under ordinary methods of cultivation would have given a precarious crop of cereals has been turned over to mixed farming; and although these farms cannot be judged on any ordinary economic basis, they are a valuable feature in the Jewish colonization as affording a livelihood for settlers and training centers for young immigrants."

In the Hills of Galilee

During the last few years the Zionist colonizing agencies, encouraged by the success of Kiriath Anavim, have established a number of farm settlements in the hills of Galilee near the Syrian border. We visited these young villages in

1939 and were particularly impressed by one of the most heroic of them, Hanita, founded in 1938 at the height of the Arab riots. It was established on 4000 dunams of land, of which 2500 were considered suitable for cultivation and the rest were to be devoted to afforestation and pasture.

Situated near the border over which terrorists constantly made their way into Palestine, Hanita had to be strongly fortified from the very beginning. It was set up in a few hours during a dark night. In addition to its eighty settlers, four hundred young people—all volunteers from towns and older settlements—came on trucks laden with wood, barbed wire and agricultural implements. They arrived at the site late at night and immediately got to work. When the Arabs of the neighboring villages awoke the next morning, they could scarcely believe their eyes. On what had always been a waste and unpopulated stretch of land there stood a wooden stockade enclosing barracks, tents and a high watchtower on which a powerful telescope and searchlight were mounted. Impressed by this display of energy, the villagers made the best of it. By noon their elders, following the custom of the land, came to congratulate the new neighbors and were cordially received with black coffee and sweets. A few days later those who had come to help in the establishment of Hanita left for their homes and the young pioneers were on their own. Each member, regardless of sex, was assigned a specific location behind the stockade in case of attack, and by day and night constant watch was kept from the high tower. This "Commando" method of establishing villages was used regularly during the dark years of 1936-39. There were, as a matter of fact, several night attacks on guards and patrols at Hanita during its early days, and the little cemetery contains the graves of three victims. But the settlers stood firm and repulsed all later attacks without loss to themselves.

There was much cleaning and grading to do before they could cultivate the land. They had to cut away thistles and dig out deeply imbedded old roots with the help of their tractor. We saw the huge roots of an ancient forest tree they had dug up in this now treeless region. The next step was the gathering up of the basalt stones scattered all over the soil. Only after all this preliminary work could the land be planted.

As we watched them planting grape vines on the newly prepared land, they told us how determined they were to make every inch of soil fertile and to cover the bare slopes with fruit trees. We are told that we could not possibly recognize the place now. More than half of the land has been planted to cereals and fodder, with patches of tobacco between them. Trucks can now use the ten-mile road from the village to the nearest highway both in winter and in summer. Milk and eggs from Hanita daily reach the markets of Haifa. The settlement is not yet self-supporting, but it is clearly on the way to full independence.

The settlers of Hanita are methodically tackling large amelioration tasks which require the work of a generation. They have already terraced some of the slopes and are directing the water of the mountain springs into channels constructed in accordance with the topography of the land. The eighty families now living in the village face the future confidently, certain that they will be able to enlarge their farm and increase their numbers.

[141]

In the Hot Jordan Valley—Dagania

Still another type of amelioration work is exemplified by Dagania. Palestine's oldest collective settlement, established in 1909, Dagania is situated in the Jordan Valley where the river leaves the Lake of Galilee on its tortuous way to the Dead Sea. Its settlers were confronted with two chief problems—malaria and inadequate rainfall. As a result of intensive cultivation of the soil and careful coverage of all rain pools, malaria has disappeared. During the first few years, fully half of the settlers were afflicted with it during harvest time and were completely incapacitated. Conditions have been so much improved that now, in spite of the very hot climate, the second generation at Dagania enjoys excellent health.

The problem of inadequate rainfall has been solved by a large pumping installation which brings water from the Jordan to all the settlement's nine hundred acres. The yields of these irrigated lands are very high and make it possible for over four hundred people to live in the two villages into which Dagania is now divided. There are, for instance, eight crops of clover a year. Every inch of soil is cultivated and Dagania is a thriving and beautiful place.

Should water from the Jordan be diverted for irrigation purposes, as proposed later in this book, hundreds of Daganias will dot the Jordan Valley. The audacious and inspired pioneers of 1909 will then have proved to be in effect the redeemers of the historic river and its banks.

The Huleh Concession

A most important reclamation project is still to be completed in the area of the concession. Huleh is the Arab name of the historic Lake of Merom, the first and smallest body of water into which the Jordan drops on its way to the Dead Sea. The valley surrounding this lake was once one of the most fertile and thickly populated regions of Palestine. Its fertility, based on a deep layer of humus, is still very great. However, as a result of the fact that the ancient drainage and irrigation systems have been neglected, pernicious swamps cover most of the Huleh area and it is the most unsanitary district in Palestine. The germs of malaria lurk everywhere. We were told that in the entire region it was difficult to find a single person not affected by the dread disease.

Before the first World War the Turkish government gave two wealthy Syrian Arabs from Beirut a concession for the exploitation of the Huleh Valley. The concessionaires paid the government $25,000 and were given a definite number of years to drain the swamps and cultivate the area. They did neither. Apparently they expected to sell their rights to Jews. As a matter of fact, in 1934 Jewish colonizing bodies organized in the Palestine Land Development Company purchased the concession and the original owners received close to a million dollars as compensation.

The area included in the Huleh concession extends over 57,750 dunams (14,500 acres). Of this area 15,772 dunams must, according to the terms of the concession, be given to the local Arab villages after the drainage work is completed. The new concessionaires plan to drain most of the 16,000 dunams of the Huleh Lake at the same time as the swamps are drained. Hence the total area of new land available for Jewish colonization will approximate 40,000 dunams or 10,000 acres.

According to preliminary estimates, every acre the Jewish

colonizing agencies reclaim will cost them about £P115, which is equivalent to $470 at the present rate of exchange. Compared to the land prices prevailing in larger and more sparsely populated countries, this is astonishingly high. But the Zionist movement, driven by the land hunger of the tragically dispossessed Jews of Europe, is willing to pay high prices for land; the initial cost is considered a national expense which should be written off the books. In the case of the Huleh, the soil is so rich that it may in the long run actually prove to be a more practical investment than exhausted tracts bought at much lower prices. Well watered by springs and streams from Mount Hermon, the Huleh Valley can yield as many as three crops a year.

If malaria is to be completely wiped out in the vicinity of the Huleh, not only the concession area but some 50,000 dunams north of it must be drained and irrigated. Part of this area has been acquired by the Jewish colonizing agencies in anticipation of future amelioration work; other parts have been included in the Huleh reclamation plans by order of the government. Thus the area to be reclaimed from waste and disease will exceed 100,000 dunams or 25,000 acres and may in time become the most fertile and thickly settled agricultural district in Palestine. Experts believe that a single family will not find it possible to cultivate more than five acres of these irrigated lands, which can produce two or three crops yearly.

The Huleh swamps were to have been drained in 1939, but because of the labor shortage caused by the war, implementation of the scheme has had to be postponed. Several initial steps have, however, been taken and an intensive health campaign is now conducted in the villages of the locality. Great quantities of kerosene are poured upon the swamps at the beginning of each malaria season to assist in annihilating the larvae of the Anopheles mosquito which is the main carrier of the disease. These preliminary efforts conducted by the Jewish health agencies and the government Health Department have already resulted in a considerable drop in the annual number of victims. When the drainage work is fully accomplished, this infested region will without doubt become as healthy as Hadera and Nahalal are today.

The Palestine Drainage and Amelioration Company founded by the Jewish concessionaires intends to begin its work by deepening the bed of the Jordan in order to divert surplus water from the surrounding area. Under the Jordan Valley reclamation program outlined elsewhere in this book, some of those surplus waters will be conducted to other districts for irrigation purposes.

Since the acquisition of the concession by the Palestine Land Development Company, several Jewish settlements have been established on the shores of the Huleh. Hulata, founded in 1935 in the southern part of the valley, is a fishing village, for Lake Huleh abounds in fish. The near-by swamps are filled with papyrus reeds from which paper may be made. According to botanists, this is the only region in Asia where this ancient plant has been preserved.

Leaching Saline Soils

Mention must be made of a smaller but most unusual reclamation project on the northern shores of the Dead Sea near the mouth of the Jordan. We came to it fully realizing its potentialities, for we had seen beautiful crops growing where only six years earlier the Dutch had drained whole

[145]

sections of the floor of the North Sea and then leached the salt out of the soil. On our visit to the forbidding-looking region of the Dead Sea we saw a group of pioneers washing the salt-impregnated soil. They do this by dividing large tracts of land into smaller boxed-in plots and keeping these completely under water pumped from the Jordan by a pipe two kilometers in length. The land is flooded for about four months. As a result of a constant head, the water percolates through the soil and leaches out the salts and carries them off with drainage waters in ditches. After the soil is freed of toxic concentrations of salt, the land is given dressings of manures and commercial fertilizers and planted to cover crops, which are plowed under as green manure, until it is ready for planting to special crops.

The climate of the Dead Sea Valley, which is more than 1200 feet below the level of the Mediterranean, is dreadfully hot. Another major inconvenience facing settlers in the region is the prevalence of sand flies which induce boils known as the "Jericho Rose." But the young pioneers, most of them German- and Austrian-born Youth Aliyah graduates, are enthusiastic and undaunted. Year by year they enlarge the cultivable area by washing additional stretches of soil and they are now producing most of their own food. They are proud of their young olive and date trees and show their visitors their Dutch-Syrian cows, which have proved able to withstand the intense heat. They cultivate tomatoes, potatoes, corn and bananas with astonishing success, and have actually begun to sell vegetables, fruit, milk and eggs in considerable quantities to the employees of Palestine Potash Ltd. as well as to the markets of Jerusalem. On eight acres of land they constructed an artificial pond for fish-breeding, and sold thousands of fish in Jerusalem last year. Their fish are fattened on natural foods washed down by the Jordan.

Inhabitants of larger and richer countries may wonder at these superhuman efforts to reclaim land. But the time may come in thickly populated and semi-starving countries, and even on our own exploited and erosion-ruined lands, when men, instead of being amazed by the efforts of Palestine's terrace-builders, swamp-drainers and soil-washers, will begin to imitate them. There are no more new continents to explore and exploit. The only new frontiers are those under our feet. What has been done in Palestine demonstrates how people in other lands, too, may achieve security and cultural satisfaction by a co-operative and scientific approach to land problems. Out of Palestine is again emerging a better way of life for tillers of the soil and a new realization of man's moral obligation to the good earth which nourishes those who give it loving care.

CHAPTER X

ARAB ECONOMY AND JEWISH SETTLEMENT
. . . .

THROUGHOUT the lands of the Near East, the control of the ruling classes has rested since time immemorial on ruthless exploitation of the peasant. Such a system is shot through with a rot that will cause it to crumble or fall when besieged by winds of adversity, or when downtrodden souls at last break forth into the red flame of revolt.

Unless the farmer has a square deal himself, he will not give a square deal to the land. If the land is neglected and exploited, it gradually ceases to produce and becomes the prey of soil erosion. As crops diminish, the farmer cannot pay his taxes and must borrow money: he can neither pay the exorbitant interest demanded nor refund the capital. He becomes undernourished, discouraged and apathetic—virtually a slave of the landowner. The "fellah" (or peasant) of the Near East could well be termed "the exploited farmer."

If the "exploited farmer" does not find it worth his while to improve the land, he will not do so. Although the natives of the Near East have fine qualities of hardihood and pride, long and ruthless exploitation and their own superstition and ignorance have kept them from realizing the full possibilities of their land and themselves. Before they can be imbued with a responsible and far-sighted attitude toward the land and its resources, the entire economy of the Near East will have to be changed and its social structure adjusted so as to permit more equable distribution of land and wealth. General education, as well as training in crafts, will have to be provided for the downtrodden peasantry, and improved agricultural and grazing methods will have to be introduced. At best this is bound to be a slow and lengthy process. It will, however, be accelerated if the fellah is brought face to face with farmers who use scientific devices and techniques and enjoy the advantages of living in a just social order. This is precisely what the Jewish settlements in Palestine can do for the fellah: their example may well be invaluable in stimulating progress throughout the Arab lands.

If we do not want the fellah to bestir himself and gain those inalienable human rights of which he knows nothing, then let us allow these Jewish settlements in Palestine to be blotted out of the Near East. But if we are interested in the regeneration of man, let all the righteous forces on earth support these settlements in Palestine as a wholesome example for the backward Near East and, indeed, for all who seek to work out a permanent adjustment of people to their lands.

The Fellah in the Near East

Fundamentally, the fellah in Palestine shares the lot of the peasants in all the other countries of the Near East with which Palestine was politically and economically united under the old Turkish regime. The semi-feudal economy still prevailing in these countries is the principal factor preventing the Arab peasant from attaining a higher standard of living. Landlords often take 55 per cent of the gross yield of the

[149]

tenant farmer while the usurer is paid according to a rate of interest that runs from 25 per cent to 100 per cent in the period between sowing and harvesting. If it is not paid punctually the interest is compounded. Furthermore, the vast majority of fellaheen own much more land than they can adequately cultivate. The fellah's farm is very inadequately equipped, and his knowledge of modern agriculture is less than rudimentary. He knows nothing of crop rotation, seed selection, manuring, summer plowing, drainage or other improvements. In spite of his low standard of living, the fellah is not able to compete with countries where modern machinery and scientific methods of agriculture are used.

The situation of the fellaheen in Iraq is very poor. In fact, even in overpopulated China I never saw conditions so bad as those I found in the underpopulated but potentially rich lands of the Tigris-Euphrates Valley. An authority on the country has described the situation thus: "The fellaheen and coolie classes were living on less than a penny a day per head. The rent of the average reed-hut, if set up on private land, varies between one shilling and one shilling and six-pence per month. . . . There are probably about two million people in the country living on such standards, and it can be imagined what purchasing power they possess, and what revenue they can offer."[1]

The rich alluvial soils and abundant waters of Iraq are practically unexploited by the present population of 3,800,000 to 4,000,000. In ancient Babylonian times, when the country was known as Mesopotamia, it was inhabited by 17 to 25 million people. Modern irrigation works could make that land maintain more people than in ancient times, but the

[1] Ernest Main, *Iraq*, p. 182. London, 1935.

masses live there in extreme poverty among their great soil resources.

The conditions of the peasants in Trans-Jordan, which was a part of Palestine before 1922, is illustrated by the fact that the local government, although relieved of certain obligations taken over partly or in whole by the government of Palestine, cannot maintain itself by taxation. According to the testimony given by Sir Arthur Wauchope, then High Commissioner of Palestine, at the twenty-seventh session of the Permanent Mandates Commission: "Trans-Jordan was a very poor country unable to pay more taxes than were at present being gathered. The Government had to be carried on; owing to the taxpayers' poverty, it could only be carried on by means of grants-in-aid."

In the course of 18,000 miles of automobile travel through the Arab lands, I was impressed by the relative progress made by the Palestinian Arabs as compared with the Arabs in neighboring countries. The British Royal Commission which visited Palestine in the winter of 1936-37, and conducted extensive hearings on the disturbances and possible remedies for them, attributed this progress to Jewish activities. In the Commission's final report, the impact of Jewish immigration on the Arab economy of Palestine is formulated in the following manner:

1) The large import of Jewish capital into Palestine has had a general fructifying effect on the economic life of the country.

2) The expansion of Arab industry and citriculture has been largely financed by the capital thus obtained.

3) Jewish example has done much to improve Arab cultivation, especially citrus.

4) Owing to Jewish development and enterprise the employ-

ment of Arab labor has increased in urban areas, particularly in the ports.

5) The reclamation and anti-malaria work undertaken in Jewish villages have benefited all Arabs in the neighborhood.

6) Institutions, founded with Jewish funds primarily to serve the National Home, have also served the Arab population. Hadassah, for example, notably at the Tuberculosis Institute at Jerusalem, admits Arab country folk to the clinics of the Rural Sick Benefit Fund and does much infant welfare work for Arab mothers.

7) The general beneficent effect of Jewish immigration on Arab welfare is illustrated by the fact that the increase in the Arab population is most marked in urban areas affected by Jewish development.[2]

The Commission report points out that "The daily wage paid an unskilled Arab laborer in Palestine is 100 to 180 mils, while skilled workers get from 250 to 600 mils a day. In Syria, the wage ranges from 67 mils in older industries to 124 mils in newer ones. Factory labor in Iraq is paid from 40 to 60 mils."[3] (A mil is one-quarter of a cent at the current rate of exchange.)

The cost of living in Palestine is somewhat higher than in neighboring countries—8 to 10 per cent above that in Syria, and 12 to 15 per cent above that in Iraq. Since wages in Palestine are more than double those in Syria and three times as high as those in Iraq, Palestinian Arabs obviously enjoy a much higher standard of living.

The higher wages paid in Palestine have attracted considerable numbers of Arab immigrants and seasonal laborers

[2] Report of the Royal Commission, p. 128, His Majesty's Stationery Office London, 1937.
[3] Ibid., p. 127.

rom other countries. According to a government estimate the
on-Jewish immigration to Palestine, both legal and illegal,
vas between 25,000 and 30,000 in the period between 1922
nd 1925. During Palestine's most prosperous years, 1933-36,
n even larger number of Arab workers from Iraq, Syria, Trans-
ordan and even the Arabian desert were employed on Pales-
ine's farms, roads and other public works. Many of them re-
urned to their homes when the protracted Arab riots brought
n the depression of 1936-39. Some of them had never in-
ended to stay in Palestine, they worked there in order to be
ble to pay debts or acquire land in their native countries.
he inflated land prices in Palestine made it difficult to
cquire a sizable plot of land there, while in Syria, Trans-
ordan or Iraq the money laborers earned in Palestine had
much larger purchasing value. The fellaheen of Palestine
self often add substantially to their meager incomes by
orking as hired laborers in near-by Jewish settlements.

Before the riots of 1936 thousands of Arab villagers were
mployed at comparatively high wages in the privately
wned Jewish citrus plantations of Judaea and the Sharon
lain. These outside earnings helped many to pay their debts
nd cultivate their own holdings more profitably. Palestinian
ellaheen have thus found it far easier to gain economic inde-
endence than the peasants of the neighboring countries
where earning opportunities are so much more limited.

Palestinian fellaheen have still another source of income—
many of them sell products of their farms in Jewish villages
nd towns. Furthermore, Arabs have always constituted the
overwhelming majority of the labor on government public
orks. On this point the Royal Commission Report remarks:
The whole range of public services has steadily developed to

[153]

the benefit of the fellaheen . . . the revenue available for those
services having been largely provided by the Jews."[4] A
minority of the population, the Jewish community has always
furnished the greater part of the Palestine government's
revenue. In 1936 J. H. Thomas, then Secretary of State for
the Colonies, discussing the six-million-pound surplus of the
Palestine government at the time, stated that this was actually
"a magnificent tribute to Jewish enterprise. In the main it is
Jewish money."

The surplus in the government treasury made possible various
types of assistance to Arab farmers. Time and again the
government was able to forgo agricultural taxes from villages
which had had bad harvests and even to lend money to the
fellaheen until the next harvest. Spurred on by the success
of co-operatives among the Jews, the government spent considerable
sums of money on organizing credit co-operatives
in Arab villages. While some of these failed because of bad
management or the dishonesty of local leaders, others have
continued to function and are helping to decrease the power
of the usurers who prey viciously on the fellaheen.

Further evidence of the beneficial effect of Jewish colonization
on the Arab community is furnished by the figures for
Arab emigration from Palestine and neighboring countries
During 1929-30, 9500 Arabs from Syria and the Lebanon
emigrated to the New World, mainly to Central and South
America—seven times as many as emigrated from Palestine
Since the Arab population of Syria and the Lebanon is only
three times as large as that of Palestine there was proportionally
twice as much emigration from those countries as from

[4] *Ibid.*, pp. 128-129.

Palestine. In 1939 the demand for visas to America from Syria and the Lebanon was about ten times that from Palestine.

The Land Question

While the majority of the fellaheen continue to employ primitive methods of land cultivation, many Arabs have been induced by the example of neighboring Jewish settlers to improve their breeds of domestic animals and to acquire better seeds for their fields and gardens. In Arab villages, near Jewish settlements, one often finds better built houses, well-planned citrus groves, imported varieties of Leghorn hens and farm stock. The improvement in farming methods is particularly marked among middle-class peasants who own their own land and have sold part of it at the high prices Jewish purchasers must pay in Palestine. The Arab owners, having cash at their disposal, can buy machinery and fertilizers and, in general, raise the standards of their agricultural work. The Royal Commission Report points out that Jews, because of the pressure of refugee needs, pay far more than the land is actually worth. I found that they paid three or four times what similar plots cost in Syria and much more than the same type of land would sell for in Southern California.

In cases where the land belongs to an absentee owner and the tenants are forced to move because the land is sold to Jewish colonizing bodies, I found that the Jewish purchasers had provided compensation to enable the tenants to lease other property. The Jews have proved to be far more considerate of the fellaheen than the Arab political leaders, who, though complaining about the alleged injury to poor fellaheen by Jewish colonization, are up in arms when any progressive agrarian reform is proposed for Palestine. Most

[155]

of them are themselves large landowners, and any change in the tenant-owner relations would not be in accordance with their interests. On the British side many officials are apparently deeply imbued with old colonial traditions, and prefer to deal with a few feudal chieftains rather than to strengthen the British position in the East by acquiring the sympathy of the masses.

In this connection, it must be said that the total area bought by Jews is comparatively small. During the last twenty-five years they have acquired only about 6 per cent of Palestine's 26,000,000 dunams: that is approximately 1,600,000 dunams or 400,000 acres. Most of this land was originally covered with swamps or rocks and considered unsuited to cultivation; less than a fourth of it was actually cultivated by Arabs.

In the three years from 1933 to 1936, when Jewish immigration reached its peak, not less than 3,702,080 Palestinian pounds, equivalent to about 18,000,000 American dollars, were paid by Jews for Arab land. What have Arabs done with the huge sums which Jewish land purchasers have put into their hands? Many, as I have previously mentioned, have modernized their own farms. The expansion of citrus production has been particularly marked: at least six times more Arab-owned land is now planted with citrus than in 1920. In the Maritime Plain Arab-owned citrus plantations cover 135,000 dunams and represent an investment of £P6,500,000. Arab capital has also gone into building and industry. Wealthy Arabs in Jerusalem and Haifa have built hundreds of comparatively modern houses and office buildings, many of which were leased to Jewish immigrants. Even in the purely Arab townships of Ramleh, Ludd and Tulkarem, situated in the vicinity of Jewish settlements, one is impressed

[156]

by the great number of new stone buildings and the general air of prosperity.

Improvement in Arab Health

An interesting sidelight on the beneficial effect of Jewish colonization on the Arab population is provided by the extraordinary rate of increase of the Arab community. From 1920 to 1940 the Arab population grew from 650,000 to about 1,050,000. Arab immigration accounts for only a very small part of this increase; most of it resulted directly from an astonishingly high birth rate and a decrease in the death rate. The accompanying table analyzes the movement of Palestine's Moslem population during the last fourteen years.

	Birth Rate Per Thousand	Death Rate	Rate of Natural Increase	Infant Mortality
1927-1929	52.7	30.2	22.6	208
1938-1940	47.0	20.3	26.7	132

The high birth rate of primitive peoples is generally counterbalanced by a high mortality rate. In the case of Palestinian Moslems the birth rate is at the primitive level, but the death rate has rapidly declined because of the modern influences at work in the country. The result is the highest rate of natural increase in the world.

Arab infant mortality is rapidly declining in all of Palestine but it is lowest in the localities closest to Jewish settlements, whether urban or rural. The reduction in the death rate is due in part to the health work of the Palestine government but must be attributed mainly to the activities of the Jewish health agencies, among which Hadassah, the American Women's

Zionist Organization, has played the outstanding part. Hadassah began its medical work during the First World War and ever since then has maintained a large network of clinics and hospitals. Its preventive medical work has been largely responsible for the elimination of trachoma, and has contributed in great measure to the effective control of malaria. Its institutions are open to Arab as well as Jewish patients. In May, 1939, while we were in Jerusalem, the magnificent structure of the Rothschild-Hadassah-University Hospital was opened on Mount Scopus. It is the most modern and best-staffed hospital in the Near East and attracts patients from all the countries around Palestine. It is making a significant contribution to the Allied war effort by arranging much-needed courses in tropical medicine for military doctors.

Students of Palestinian Arab life agree, and my own observations bear them out, that the villages near Jewish settlements are much less backward than those in all-Arab sections. In the latter the houses are generally small clay huts with few, if any, windows; these are shared by the fellah's family and his farm animals—an emaciated donkey, cow and pair of oxen. The fellah's only farm implement is a wooden plow. In and around the village there are usually few trees. Around the village lie heaps of dung which have accumulated in the course of generations. The fellah's children run about half-naked and barefoot. Many of them suffer from trachoma and become blind. There are no sanitary arrangements even of the most elementary kind.

In contrast to these primitive conditions we found that Arab villages in the neighborhood of Jewish colonies show signs of genuine progress. In many of them the fellaheen live in comfortable little houses of brick or stone with red-tiled roofs and separate accommodations for cattle and other domestic animals. The streets in these villages are broader and more sanitary than they were a generation ago. The children, still barefoot, look cleaner and healthier. Eye diseases and malaria are far less prevalent than they used to be. Agricultural methods have been considerably modernized as a result of Arab imitation of Jewish neighbors. Indeed, near Jewish villages we often saw modern orange groves and irrigated vegetable gardens belonging to Arabs but almost indistinguishable from those of Jews.

A case in point is the village of Zarmuka adjacent to the prosperous Jewish settlement of Rehovoth. About half of Zarmuka's 13,000 dunams of land were sold to Jews, and the villagers used the money they received to plant orange groves on much of their remaining land. Before Zarmuka sold part of its land, it had eighty dunams of orange groves; now it has approximately 1800. In addition, the villagers cultivate large irrigated tracts of vegetables sold chiefly in near-by Jewish villages. The fellaheen of Zarmuka own European plows, harrows, threshing machines and other agricultural implements, all of which Jewish farmers have taught them to use.

In the cities, too, the standard of living of the Arab masses has risen markedly during these recent years of progress. We found that the average Arab in a Palestinian town consumes better food and looks more prosperous and healthy than the average inhabitant of Trans-Jordan and Iraq, and even of so rich a country as Egypt.

That the Arab population is most prosperous in districts where Jewish activities proceed on a comparatively large scale is demonstrated by government statistics. The table compares the population of a number of towns in 1941 with their population in 1922 and 1931. The towns are divided into three categories: (1) those with a mixed population of Arabs and Jews; (2) predominantly Arab towns in the vicinity of Jewish settlements; and (3) Arab towns in purely Arab districts.

A. Towns with Mixed Arab-Jewish Population

Name of Town	No. of Inhabitants			Increase	Increase
	1922 (census)	1931 (census)	1941 (estimate)	1931 over 1922 %	1941 over 1922 %
Haifa	24,634	50,403	113,000	104	354
Jerusalem	62,578	90,503	138,000	45	121

B. Arab Towns in Vicinity of Jewish Settlements

Jaffa	32,524	51,866	84,000	59	158
Ramleh	7,312	10,421	13,300	43	82
Lydda	8,103	11,250	15,500	39	91

C. Arab Towns in Purely Arab Districts

Nablus	15,947	17,189	21,600	8	35
Gaza	17,480	17,046	21,500	minus 3	20
Bethlehem	6,658	6,815	7,800	2	18
Hebron	16,577	17,531	20,600	6	24

Since in the period between 1922 and 1941 the Arab population of Palestine increased from 650,000 to above a million, in a district where no immigration or emigration took place the Arab population of 1941 should exceed that of 1922 by 55 per cent. The increase of population in the towns in the purely Arab districts of Palestine was, however, much less. It amounted to 35 per cent in Nablus, 24 per cent in Hebron, 23 per cent in Gaza and only 18 per cent in Bethlehem. Evidently there must have been inner migration in Palestine from the purely Arab districts into those with a considerable

Jewish population. The table makes it very clear that though Arab leaders demanded stoppage of Jewish immigration on the grounds that it was harmful to the Arab masses, the masses actually left their own towns to settle in cities and villages where there was more economic contact with Jewish immigrants.

Two Arab Claims

Our review of the impact of Jewish settlement on Arab economy will not be complete unless we analyze two statements often made by Arab opponents of Zionism. One is the assertion that Jewish land purchase has left many Arabs landless; the second is the complaint that Jews do not employ Arabs on their farms and in their factories.

A good test of the truth of the first statement was made in 1931 when the government of Palestine announced its readiness to finance the resettlement of all Arabs able to prove that they had become landless as a result of Jewish land purchase. Nationalist leaders toured the villages and towns urging Arabs to file claims. Over three thousand of these were registered with the government but the greater part were quickly rejected as utterly unfounded. The claim of 656 were found to be warranted. But fully half of the 656 Arabs involved rejected the offer of resettlement, preferring to remain in the cities where they had found steady employment and better living conditions.

That this situation continued to hold true in Palestine is demonstrated by a statement in the report made by the High Commissioner for Palestine to the League of Nations in 1935:

All registered Arabs who have signified their willingness to take up holdings upon Government estates have been accommodated.

[161]

That only a small number of Arabs have come forward so far to take up land is due to the fact that there is at present plenty of employment to be found in the towns and neighboring groves.

The assertion that Jews as a rule do not employ Arabs on their farms and in their factories deserves detailed consideration. During my visits to Jewish co-operative and collective villages I saw no Arab workers, although I did see a considerable number on privately owned Jewish farms. When I asked the members of a co-operative village whether they accepted Arab labor, they told me that in their case the question was purely theoretical. All these so-called labor settlements are based on the principle of what they call "self-labor": in other words, all work is done by the members of the settlement and no outside laborers, whether Jewish or Arab, are employed.

In industry there is an undeniable tendency to employ Jews in factories built by Jewish capital, and Arabs in factories belonging to Arabs. We found, however, that Jewish owners willingly employ Arab as well as Jewish labor in cases where the undertaking serves both sections of the population. Two outstanding instances of this are the Palestine Electric Corporation and the Potash Syndicate, both organized and managed by Jews. The latter in particular employs Arabs in great numbers. In public works, harbors, etc., Arabs are always the majority of the workers, though Jews pay much more of the taxes which make these projects possible.

The leaders of the Jewish labor movement in Palestine are genuinely sympathetic with Arab labor, but believe that Arab workers should, for the time being, be employed not in the Jewish but in the Arab and governmental sectors of Palestine's economy. They point to the economic benefits which

Jewish activities have brought the Arab population and are ready to help the Arabs modernize their agriculture and industry. The Jewish labor leaders insist, however, that since Jewish work in Palestine aims to create a Jewish Commonwealth for the persecuted masses of European Jewry, all hopes for such a commonwealth would be frustrated if Palestine were built by Jewish capital but predominantly Arab labor.

To Palestine's Jewish labor leaders, the development of two parallel economic structures seems the best way to assure the general prosperity of the country. When Arab workers become accustomed to unions and the Arab standard of living is raised to higher levels, these parallel economies will, they feel, gradually merge with each other. It is with this general goal in mind that Jewish labor unions co-operate with, and aid, groups of Arabs who wish to organize themselves and improve their working conditions.

It seems most plausible that a post-war influx of Jewish refugees will further improve the economic situation of the Arabs in Palestine. An increased number of Arabs who have land but little or no capital will be able to sell part of their land at high prices and thus acquire capital for the irrigation and general reclamation of the remaining part. Moreover, Jewish resettlement work in Palestine has never been an imperialist movement ruthlessly exploiting the native population in order that small groups of foreign investors might profit. Our observations in Palestine convince us that Jewish settlement not only has done no harm to the Arabs but has actually raised their status far above that of the Arabs in the neighboring lands.

Why, then, it is most logical to ask, have Arab leaders been

[163]

so vehemently opposed to Jewish settlement? Why have they encouraged and led riots which time and again have hampered the progress of Palestine?

These questions assume that national groups always act rationally. Nothing is farther from the truth. Even a people with a background of education and civilization may frequently act under the influence of irrational fears, blind enmity and deep-rooted prejudices. No one would claim that in following their dictator leaders the populations of the Axis countries are acting in accordance with their real interests. The Arabs can hardly be expected to show more political wisdom than the Germans or Japanese.

The fact that there is no basic economic antagonism between Jews and Arabs, and that Jewish activities have actually benefited the less progressive Arabs, gives us hope for the future. In a new world order based on genuinely democratic and co-operative principles, the Arab national movement may well be influenced to express itself more constructively. Palestine is after all only a very small part of the large area sparsely occupied by the Arab race. Arab leaders in the future may find it to their advantage to utilize the constructive example and the assistance of Jewish Palestine in raising the entire Near East to a higher level of prosperity and culture.

There was a time in the recent history of Arab-Jewish relations when influential Arab leaders expressed their sympathy with Zionism and their willingness to co-operate with the Jews in the upbuilding of the East. This was immediately following the Armistice of 1918 when under the idealistic leadership of Woodrow Wilson well-intentioned men everywhere sincerely believed that we were entering a new world of true international co-operation. In January, 1919, the late King Feisal, who had taken the leading part in the Arab desert rebellion of 1916-18 and later became ruler of the newly created state of Iraq, collaborated with Dr. Weizmann, the President of the World Zionist Organization, in issuing a declaration of friendship in which both leaders state that they are mindful of the racial and ancient bonds existing between the Arabs and the Jewish people and realize that the surest means of working out the consummation of these national aspirations is through the closest possible collaboration in the development of the Arab State and Palestine.

Feisal declared that: "The Arabs, especially the educated among us, look with deepest sympathy on the Zionist movement . . . we will offer the Jews a hearty welcome home. . . . Interested parties have been enabled to make capital out of what they call our differences. . . . I wish to give you my firm conviction that these differences are not questions of principle but matters of detail . . . and are easily dispelled by mutual good-will."

Unfortunately, the Arab movement after the war of 1914-18 did not develop in accordance with the spirit of Feisal's promises. The underlying cause for this was the rise of narrow nationalism in the world at large. It was only natural that Arab leaders should have been influenced by the general upsurge of isolationism and reactionary nationalism. Genuine co-operation among Jews and Arabs was hardly possible in a world where there was no such co-operation between Rumanians and Hungarians or Germans and Poles.

Another factor which contributed greatly to Arab-Jewish strife was the attitude of several great powers who were interested in promoting disunity in Palestine. In the ten years

[165]

preceding the present war, Fascist Italy and Nazi Germany were very active in fomenting Arab discontent. They played no small part in the prolonged Arab outbreaks of 1936-38.

As I pondered on the beneficial changes being wrought in the long-neglected Holy Land by Jewish reclamation work, I often thought of what would happen to the Jews of Palestine and to the country as a whole if Jewish immigration were effectively stopped and the land placed under full Arab control as envisaged in somewhat nebulous form by the British White Paper of May, 1939. Would the Jews be able to live securely as a substantial minority under Arab rule? Or would such rule result in constant persecution and pogroms?

The Grand Mufti, recognized leader of the Arab extremists, who is now attached to Hitler's headquarters, answered the question in his testimony before the Royal Commission in 1937. He stated that if the Arabs gained control of Palestine, all Jewish immigration would be prohibited. He added that the country would not even be able to "assimilate and digest the Jews already there." Urged to explain whether his statement meant that some of them would have to be removed by a process "kindly or painful as the case might be," he replied, "We must leave all of this to the future." The members of the Commission must have thought, as I often have, of the fate of the Assyrians, the Christian minority in Iraq. When the British relinquished their mandate over Iraq, that country's leaders promised solemnly to protect the Assyrians. Instead, the Assyrian Christians were slaughtered by Arabs of the Mufti's ilk who did not wish to "assimilate or digest them"!

Arab rule in Palestine would, I feel sure, put an abrupt end to the reclamation work now being carried on so splendidly. Erosion would begin to have its way again in the fields. Peasant women in search of fuel and goats in search of pasture would make short work of the young forests.

These statements are by no means intended to be disparaging to the Arabs. The average fellah is no less intelligent than the average peasant in backward countries, but for the time being the Arab national movement is monopolized by feudal landowners of Fascist sympathies and beliefs. These leaders are not only not concerned with, but actually antagonistic to, the social and economic advancement of the masses of their people. It seems to me that without the Jews as catalytic agent, marked improvement in the position of the Arab masses would be much slower.

All the Zionist leaders with whom I talked, eagerly anticipated the day when they would be able to co-operate with an Arab leadership which, while defending the legitimate interests of its people, would reject the Fascist philosophy and renounce bloodshed, arson and violence as political techniques. I found that Jewish leaders, without exception, were deeply interested in the social welfare of the cruelly exploited Arab workers and fellaheen. They recognize that any sore spot in a country's economy affects the entire population and they wish to help their half-brothers, the Arabs, reach a higher level of progress and prosperity.

In my opinion, the prospects for Arab-Jewish understanding after the war depend less on the two peoples themselves than on the general climate of post-war international relations. Only if the victorious United Nations co-operate sincerely in the building of a free and humane post-war world will it be possible to settle national conflicts justly. The Arab-Jewish situation is no exception to this rule.

[167]

CHAPTER XI

THE JORDAN VALLEY AUTHORITY—A COUNTERPART OF TVA IN PALESTINE

. . . .

As A LAND conservationist, forestry engineer and hydrologist, I have had the opportunity to examine and study many of the world's greatest reclamation projects. I have visited and studied the dykes by means of which the Chinese have for six hundred years maintained the Yellow River within ever mounting levees across four hundred miles of its delta plain. I have seen how the Dutch have made fertile farm land of 550,000 acres on the floor of the Zuider Zee; how in southwestern France 440,000 acres of sand dunes have been changed from a rolling menace into productive and life-giving forests and 2,500,000 acres of marsh land have been drained to provide homes for hundreds of thousands of people. In the French Alps I have seen remarkable torrent correction and flood-control works, and in Italy I have examined the reclamation of the Pontine Marshes. Here in the United States I have studied the Tennessee Valley Authority; the Boulder Dam, with its 250 miles of huge conduits that supply power and water for our Southwest; the Grand Coulee power and irrigation project in the Northwest; and other important undertakings. These vast projects demonstrate how modern engineering can harness wild waters to produce cheap power

[168]

for industry and how scientific agriculture can transform waste lands into fields, orchards and gardens supporting populous and thriving communities.

The Valley of the Jordan River offers a combination of natural features and a concentration of resources which set the stage for one of the greatest and most far-reaching reclamation projects on earth, comparable to our TVA in scope and importance. In 1939, while making an airplane survey of Palestine, I looked into the deep rift of the Jordan Valley which drops to 1300 feet below sea level and lies only a short distance from the Mediterranean Sea, and I realized that this extraordinary difference in altitudes offers a splendid opportunity for a great power project. Later I was told that about fifty years ago a French engineer had rather vaguely proposed something of the sort and that there is also some mention of the idea in the writings of Dr. Theodor Herzl, the founder of modern Zionism. Half a century ago, however, such a suggestion could not have been taken seriously. Palestine was then in a state of utter backwardness and Jewish settlement was just beginning. Nor were engineers nearly as experienced as they now are in planning and executing large reclamation projects.

Further study of the possibilities of what I shall call the Jordan Valley Authority or JVA, has convinced me that full utilization of the Jordan Valley depression and adjoining drainage areas for reclamation and power will in time provide farms, industry and security for at least four million Jewish refugees from Europe, in addition to the 1,800,000 Arabs and Jews already in Palestine and Trans-Jordan. The rehabilitation of the European Jews who will survive Hitler's infamous regime will be one of the greatest problems the United Nations will have to face after the war. Indeed, it is a debt of honor owed by the Christian peoples to those who have suffered most in this world-wide struggle for freedom. The JVA, building on the remarkable achievements already accomplished by Jewish settlers in Palestine, has a great contribution to make to the solution of the Jewish refugee problem and can at the same time serve as an example for all the other lands of the Near East.

The Aims of the JVA

Palestine has two primary needs: water and power. Water is available in the flow of the Jordan and potential power is locked in the swift and turbulent descent of the river to the depth of the Dead Sea. The main aims of the JVA are thus the diversion of the sweet waters of the Jordan and its tributaries for the purpose of irrigating the arid lands of the Jordan Valley and its slopes, and the utilization of the deep incline of the Jordan River channel for purposes of power development.

The irrigation program of the JVA envisages the diversion of the sweet waters of the Upper Jordan and of the Yarmuk and Zerqa rivers into open canals or closed conduits running around the slopes of the Jordan Valley. This would provide water for the irrigation of the slopes and floor of the Jordan Valley. Alkaline soils would be leached out by these fresh waters and made suitable for farm forage crops. On the basis of three acre feet of water required for year-long irrigation in the Jordan Valley, the total area that can be irrigated by the fresh waters of the Jordan drainage amounts to 300,000 acres of land. As this may exceed the area of irrigable land in the Jordan Valley, which is about 155,000 acres, there will be a surplus of water available. This surplus will have to be taken out above the Huleh at about 600 feet above sea level and will flow by gravity to provide for irrigation in the Plains of Esdraelon, Beisan and possibly some small valleys of Galilee en route.

In addition to the irrigation plan the JVA calls for the development of power facilities. In order to understand this power program it is necessary to review the unique topographical features of the Jordan Valley. This rift valley has created the greatest inland deep on the face of the earth. Its total depth is 2600 feet below sea level but it has been filled in by the Dead Sea up to 1300 feet. Owing to this low level, the Dead Sea has a dry, torrid climate and its annual evaporation is estimated by Donder at six to sixteen feet of water per year[1] and by Press at seventeen feet per year.[2]

The power program calls for the introduction of sea water from the Mediterranean into the Jordan River Valley for the double purpose of compensating the Dead Sea for the loss of the diverted sweet waters of the Jordan and of utilizing the sea water for development of power. The Jordan Valley is about twenty-five miles from the Mediterranean at Haifa Bay. Of the three technical possibilities of effecting the link, the JVA plans provide for an open canal of some seven miles from near Haifa to Mount Carmel and for a twenty-mile tunnel through the Plain of Esdraelon to the edge of the great chasm of the Jordan Valley. The canal and tunnel would be designed to carry enough sea water—about one thousand cubic feet per second—to make up for the annual diversion

[1] M. J. Ionides, *Report on the Water Resources of Trans-Jordan*, p. 143, 1939.

[2] Reported by Ch. Audebeau Bey, Ministère de l'Agriculture Direction des eaux et du génie rural, *Rapports et notes techniques*, fasc. 57, 413, 1927.

of sweet water from the Jordan, and thus maintain the Dead Sea at its present level. The introduction of sea water into the Jordan Valley might cause some damage to adjoining lands because of seepage. This contingency can be met by conducting the sea water into the Dead Sea in concrete-lined canals; the construction of such canals may have to be undertaken in any case, because of the unsuitability of the Jordan River bed as foundation for dams. As this sea water dropped into the Jordan rift, there would be something less than 1200 feet of effective fall for the development of hydroelectric power in a number of power plants. It is estimated that 1000 to 1500 second feet of water dropping through 1200 feet would have a minimum capacity of 76,000 to 110,000 of kilowatts with an installed capacity of roughly twice these amounts. Additional amounts of power generated by other developments in connection with the projected irrigation schemes would bring the total to about 150,000 kilowatts. This capacity, exceeding that of the Norris Dam, would yield approximately 1,000,000,000 of kilowatt hours per year, an amount of electrical energy which, under Palestinian climatic and other living conditions, would be sufficient to serve the needs of well over a million of additional population. The supply of water and water power could be further increased by the utilization of water resources which lie in areas adjacent to Palestine and which are now not being utilized. There is no doubt that a regional approach to the problem of water resources based on agreement of all potential interests involved would greatly enhance the economic significance of the project and redound to the benefit of all concerned. It should be clear that a plan which harbors such vast economic potentialities merits the immediate utmost attention of all interested in the post-war problem of Jews and Palestine.

Power and irrigation are only part of the JVA project. Its general objective is the development of the land, water and mineral resources of the Jordan drainage in Palestine and Trans-Jordan, and the maritime slopes of Palestine. As in the case of the TVA, water conservation and flood control would be an important part of the JVA's activities. Engineers would be called on to find appropriate locations for and construct check dams and dams for reservoirs, both to impound water and to sink surface flow into underground drainage, as well as to divert water for spreading on pasture and farm-crop lands. These water conservation measures would be made to fit into a plan for flood control to protect lower-lying works, farms and homes.

The JVA would also seek to promote conservation of land by controlling soil erosion and introducing measures to increase the soil's absorption of rainfall and to lead unabsorbed waters away from fields and terraces to natural drainage channels without cutting and carrying away the soil. The repair, adaptation and improvement of old conservation measures would be supplemented by the development of new ones. The problems of soil improvement, increase of organic content and fertilization would be included in this program.

The important but long-neglected problem of scientific range management and grazing would also fall within the jurisdiction of the JVA, for a very large part of the total watershed area is best suited for grazing. Attention would be given to the sustained maximum carrying capacity of the range and to the improvement of livestock. Direct returns in greater yields from grazing lands can be expected within five years.

[173]

Under the JVA's direction, moreover, measures would be taken to make the economies of grazing and farming supplement, rather than conflict with each other. There is urgent need in the Near East for a demonstration on a small scale, such as could be carried out within the area of the JVA to solve the age long antagonism between the shepherd and the farmer, which has repeatedly brought about the destruction of highly developed cultures and civilizations. When these two economies have been scientifically worked out as complementary instead of antagonistic, then not only will this JVA solve one of the great points of conflict throughout the centuries, but will provide an example which can in time, be applied to bring about increased prosperity and security for the Near East as a whole.

The reforestation of lands unsuited to farming and grazing would also be undertaken by the JVA: reforestation will provide much-needed fuel and timber and restore a more regular regimen of storm runoff water. The millions of trees already planted by the Jewish colonizing agencies and the Palestine government demonstrate that forests can grow as in ancient times. The beneficial effects of a reforestation program will be felt within as short a time as ten years.

An important and lucrative phase of the JVA would be the extraction of important minerals from the Dead Sea waters on a scale far larger than the present one. Potassium for fertilizers and war industries, bromine for oil refineries, magnesium for alloys utilized in airplane construction—all await exploitation in vast quantities.

The full-scale draining of the Huleh, discussed above, would be co-ordinated with the activities of the JVA, as would the reclamation of the fertile, alluvial lands covering about 25,000 acres in the vicinity of the Huleh.

Also included within the scope of the JVA would be the reclamation of the Negeb, or South Country, which comprises an area almost equal to that of the rest of Palestine. The cheap power available under the JVA would enable the decentralization of thriving industries into this region. Furthermore, the extensive areas around Beersheba should be developed by irrigation. The JVA would provide funds for thorough investigation of artesian water supplies, and for the construction of dams for storing rain water from the Hebron Dome and conducting it to the Negeb, where it would be used for irrigation purposes. During the rainy season in Palestine, we saw the flood waters of the Yarkon and lesser streams roaring out to the Mediterranean. Under the JVA these waters, too, would be utilized. The JVA would construct dams higher up in the hills to hold excess waters temporarily until they could be conducted by canal and conduit to Beersheba and other fertile but waterless areas in the South Country. It might be necessary to construct one pumping station, but this would be a relatively simple matter; oil, it will be remembered, is pumped six hundred miles from Mosul in Iraq to Haifa and must even go over a mountain pass in the Lebanon.

The practicality of using flood waters to irrigate the Negeb is confirmed by successful American experiences in Southern California. In general, comparison between the JVA and other great power projects makes it clear that the JVA would not have to contend with many of the difficulties faced by other projects. Its engineering problems would be far less difficult than those at Boulder and Coulee dams. It may be necessary to acquire some inhabited farm land for reservoir sites, but to a far lesser extent than was required by the TVA. There would also have to be some adjustment with economic groups now operating in this area. The Palestine Electric Corporation owns one hydroelectric station located on the Jordan and oil-burning power plants in Tel Aviv and Haifa. This company also has the concession for the production and sale of power for all of Palestine except the Jerusalem district, which is supplied by an independent company. Similarly, the Palestine Potash Company owns a concession for the exploitation of the minerals of the Dead Sea. Suitable arrangements would have to be made with the interested groups so as to bring them within the larger scope of the JVA.

What will the JVA cost in dollars and cents? It is obviously difficult to give a precise answer to this question, especially at a time of war and uncertain purchasing power of the money unit. It may be said, however, that a preliminary study of geographical conditions in Palestine does not reveal any great technical difficulties and justifies the expectation that costs will not exceed those involved in other projects of this kind. On the contrary, as the maximum elevation between the Haifa Bay and the Valley of Beth Shean, where the proposed canal and tunnel are to be built, does not exceed 150 feet, the basic work of the JVA may even cost less than similar other undertakings.

Moreover, vast quantities of needful equipment will be available near by. The Near East is now a great military camp crowded with supplies and equipment. After the war, military equipment may be converted to works of peace—a modern version of "beating swords into plowshares." Military tanks can be changed over to bulldozers for necessary excavations; and military trucks may be used for the hauling of materials, rock and earth. Labor can be supplied by local people and more especially by new refugees moved out of concentration camps in Europe. Materials for excavation and construction will likewise be ready at hand; explosives for tunneling and construction, and cement from enlarged plants such as exist now at Haifa. Not only will there be the need for the JVA project, but equipment, labor and materials will be ready at hand or quickly converted to peaceful production.

How can a project of such vast proportions be financed? Both the magnitude of the task and the urgent needs the project must serve point to public international sponsorship of the JVA. It is thus proposed that the United Nations and public Jewish organizations participate in the financing of the project. It is possible also to envisage the participation of private investment capital; the JVA might well take the not uncommon form of quasi-public authority operating with both public and private capital. Many Jews will want to participate in this project; and millions of Christians would wish to share in the rejuvenation of the Holy Land by contributing to the reforestation of the barren hills and other good works. The Jews, who for centuries have been so cruelly persecuted at the hands of Christians, should be made custodians of this new Holy Land and directors of the JVA under the supervision of the United Nations.

It is also proper to inquire: Will the project "pay"? Will the material benefits yielded by the development be commensurate with the vast outlay of funds? Here, too, precise advance calculation is not possible. Like the TVA, the Jordan development is a multi-purpose enterprise. It consists of many component tasks. Some of the activities, such as the

[177]

increased extraction of minerals from the Dead Sea, should bring direct and immediate revenue. Returns from the sale of electricity will likewise be immediate and large. Returns from irrigated lands may be expected to be high, especially since crops in the tropical Jordan Valley can be grown practically the year around. Other activities, such as improvement of soil, development of grazing lands and reforestation, will in time yield increasing direct and indirect returns. The JVA project should be considered as a whole: immediate returns should be applied to defraying the cost of those activities in which income would be delayed or indirect. Eventually the JVA, managed as a national enterprise, is expected to be self-liquidating.

What of the million and a third Arabs in Palestine and Trans-Jordan? They would benefit greatly from the JVA. The increased Jewish immigration it would make possible would enlarge the market for their produce and provide them with new opportunities for investment and labor. If individual Arabs found that they disliked living in an industrialized land, they could easily settle in the great alluvial plain of the Tigris and Euphrates Valley where there is land enough for vast numbers of immigrants. The soil of Iraq is so fertile and irrigation waters are so abundant in the Tigris and Euphrates rivers that the land needs more farmers. I was present at the opening of the Kut Barrage or Diversion Dam, the first built on the Tigris River, which has put into cultivation a considerable area of the ancient farm lands. Iraqi officials told me then that there were not enough farmers in the country to make use of the water that could be diverted from this one dam. There is land enough and great possibilities for land restoration in the Arab countries; they can be brought to far greater productivity and prosperity than they now enjoy. Centuries of neglect of their lands and consequent wastage demonstrate that the Arabs have not shown the genius or ability to restore the Holy Lands to their possibilities. But the Jews, by their magnificent examples in colonization, have demonstrated their ability to reclaim and redeem wasted lands from desolation to high productivity. Moreover, in the case of the Jews there is no place except Palestine where additional Jewish refugees can be placed at the close of this war and be rehabilitated physically, spiritually and economically after experiencing a martyrdom unparalleled in world history. By making possible the settlement of millions of Jewish refugees in their historic homeland, we shall wipe out one of the darkest blots on our civilization—the persecution of the people who have given us the foundation of our religious conception and the basis of the democracy for which we fight today unto death.

Essential technical features of the Jordan Valley project have not been included within this chapter. But they have been worked out and submitted to a number of the foremost consulting engineers of America, who have pronounced the project as feasible and workable and capable of being constructed at no greater cost than similar irrigation and power projects in the western part of the United States.

CHAPTER XII

THE POSSIBILITIES OF THE NEGEB

. . . .

THE Negeb, as the triangular stretch of land comprising the southern portion of Palestine is called, consists of three million acres or almost half the area of Mandated Palestine. It is an undulating plain that fans out from the elevated Hebron Dome eastward to the Dead Sea and down to Aqaba on the right arm of the Red Sea. From Gaza it extends thirty miles to the south along the Mediterranean, reaching the Egyptian border.

We first visited the Negeb in February, 1939, on our automobile journey through the desert from Egypt into Palestine. A dirt road, often covered by sand and camel footprints, led us across the brown landscape of Sinai. It followed along wadies in which isolated acacia trees grew, browsed as high as goats could stand or climb, and it climbed over saddles of undulating sunburned hills. After a night at the Egyptian Patrol Rest House we crossed into the "South Country" or the Negeb.

At the border, we saw an example of the age-old struggle between the people of settled abode and the herdsmen, which has brought destruction to the Holy Land time after time from the beginning of history. Houses were wrecked, telephone wires were dangling and in some places there was evidence of destruction by fire. Former occupants had been killed or had fled, and their cultivated gardens were dying from lack of irrigation. We startled a Bedouin clan who had pitched their black goat's-hair tents just outside the group of wrecked and looted houses; the Bedouins stared ominously, and their ferocious dogs charged at us. Donkeys, camels and sheep browsed on former gardens and trees. It was an excellent example of the manner in which marauding nomads from the desert sweep down upon a settled population, killing or dispersing the settlers and destroying the works of their hands.

There were no Jewish settlements in the part of the Negeb which we traversed on our journey to Palestine, but the desert tribesmen, originally incited by the slogan of protecting Islam from "Jewish invaders," used the opportunity to loot and plunder the settlers at the border though they spoke the same language and professed the same religion. This was in accordance with old customs and precedents: kinship and religion are forgotten when the nomads of the desert arise to continue the old feud of Cain and Abel.

On the hill overlooking the wrecked border post and the encampment of Bedouins stood the ruins of a monastery where an unusually interesting manuscript of Byzantine times had been found. Throughout the "South Country" to Beersheba we came on similar contrasts between the ruins of towns and the flocks of roving nomads who pitched their tents wherever water and a few blades of grass could be found. We were to see many more such scenes during our later field studies and airplane survey of the Negeb.

Aside from the coastal plain, the Negeb is barren and sunscorched. The only evidences of vegetation and culture are an ephemeral covering of vegetation after winter rains, and brush growth, aromatic tamarix and thorny acacias along wadi courses and old rock walls built long ago across shallow valleys to hold back the soil. The average rainfall diminishes from about fourteen inches at Gaza to eight at Beersheba, and reaches the minimum of three inches at Kurnub. In the southwestern part of the Negeb we experienced a sandstorm and watched sand dunes form and roll inland. It was a weird demonstration of one of the forces which contribute to the creation of "man-made deserts."

There is no archaeological evidence to prove that the Negeb was populated in early Bible times as was the fertile crescent further to the north. But in the ruins of six cities, and of numerous smaller places, dams, cisterns, terraces and works for the conservation of soil and water, we find evidence that a highly developed and energetic people flourished in the Negeb in post-Biblical times. None of the terrace walls appear to have been built earlier than the Greek period, and the prosperity of the Negeb reached its peak under the Byzantine Emperor Justinian in the sixth century. According to archaeological evidence, civilization in the Negeb began with the Nabateans, a people of Arab origin who settled there in the later centuries of Roman rule.

The Nabateans accumulated much wealth because of their favored position at the crossroads of caravan routes between Africa and Syria and between India and Mediterranean ports. They were thus able to build towns along the caravan routes, and to undertake remarkable measures for soil and water conservation. No other ancient people worked out the refinements of soil and water management to such a high degree as to support agriculture in an arid, difficult region.

[182]

THE POSSIBILITIES OF THE NEGEB

My studies at Petra in the southern part of Trans-Jordan demonstrated that the Nabateans, though indeed energetic traders, also went to great lengths to develop agriculture to support their cities.

During Byzantine times Christian monasteries flourished in the Negeb where they found "uncontaminated space" for hermitages. A hamlet or a village arose around each. Written records of the times and recitals of pilgrims tell us of a great and holy population in this southern desert region.

The Romans gave Palestine a long period of peace. The population multiplied and developed the country to a remarkable degree. Under the compulsion of supporting a large population, reclamation works were carried out; swamps were drained; aqueducts were built, even with the use of inverted siphons in Trans-Jordan; dams were constructed; and the whole country was served with roads that crossed over stone bridges at almost every stream. Farmers replaced their mud huts by stone buildings; hamlets sprang up in the wilderness; villages grew into towns planned on the Roman model; houses had porticos, and colonnades were built along streets leading to magnificent marble temples, luxurious baths and fine private houses. From the Euphrates to the Red Sea the ruins of this period transcend those of earlier times and bear witness to a population more numerous and wealthy than the land had seen before or since. It was during these times that people pushed out into less habitable areas and, by ingenuity born of need, devised ways to make effective use of the meager and erratic water supply offered by sudden winter rainstorms.

Woolley and Lawrence, who made a study of this region, state:

The Negeb is emphatically a country either for nomads whose camels and goats may contrive to exist upon the scanty pasturage of the stunted scrub, or else for a very clever and frugal agricultural people who can husband such little water as there is, and overcome the niggardliness of nature. In no respect is it a land for a large population, and the considerable towns whose ruins now surprise us in the waste, obviously owed their existence to extraneous forces. It is, we think, both natural and correct to assume that at all periods in man's history, the southern desert has been very much the desert that it is today.[1]

It is apparent that the Byzantines relied solely upon stored water: they showed extraordinary ingenuity and skill in developing water in wells, storing rain water in cisterns and diverting runoff waters into the fields on valley floors along the wadies. Woolley and Lawrence comment on this point:

The elaborate terracing system of the Byzantines, which turned to account the natural infiltration of the rain water through the earth, was the main secret of their agricultural success. Hedges of tamarisk were planted along terrace walls and around fields at once to bind the light soil, to break the force of the winds and to attract moisture. We noticed that wherever these terrace walls are preserved, and especially if their hedges yet remain, on these the modern Bedouin prefers to sow his corn and crops, and the crop is in the best condition. The use of tamarisk for hedge rows is perhaps one more sign of the prevailing drought of the country, for it is one of the few shrubs which without irrigation, will withstand the heat of summer and autumn in the desert today, and probably that same hardihood recommended it for the same climate in Byzantine times.[2]

[1] C. Leonard Woolley and T. E. Lawrence, *The Wilderness of Zin*, Palestine Exploration Fund, 1914-15, p. 17.
[2] *Op. cit.*

Byzantine treatment of the water problem seems to have advanced along fairly scientific lines, according to Zissu and Gwyer. There is good reason to suppose that before the end of the Byzantine period, their engineers were acquiring the skill required for tapping the underground water supply. At Ruheiba, there is a well over three hundred feet deep and apparently of late Byzantine construction. It was cleaned out a few years ago and water was found at its original level.

The ancient cities and ruins of this region have been described by Woolley and Lawrence, by Guy, by Jarvis and by other archaeologists and students of this fascinating area. They tell of an entirely different type of occupation from that of the Bedouin, who leaves few enduring marks of human activities on the land.

South of the line on the map from Auja to Beersheba you come to remains of an intensive Roman settlement in which there were six large towns that accommodated from five to ten thousand people each—namely, Auja and Esbeita, north of that Khalasa, then Ruheiba, and southwards Abda and Kurnub.

The people who lived in those towns did not rely on the wells because they were few and on the whole unsatisfactory. The water is rather saline. They relied almost entirely on the rainfall and collected it by a variety of cisterns. Every house had a cistern in the basement, as they have now in Jerusalem. In every town also there were enormous cement reservoirs in which the whole of the rain was collected, and dotted all over the mountainside were these underground cisterns or harabas carved out of the living rock. It is very difficult to say how many there are. In Sinai I knew of the presence of six or seven, but when I started to clean them out and plant olive trees in their vicinity the Arabs pointed out many more. They are entirely filled with silt, and there is nothing to show where they are beyond the cut-stone in a heap on the surface, and this is now so weathered as to be difficult to detect. I should say there are at least a hundred of those harabas in this area. When I tell you that most of these harabas are about the size of a large room and the same depth, and they have all been carved out by small hand-tools, you will realize that there must have been a very great necessity for them and the fullest use was made of them.[3]

Woolley and Lawrence report village sites strewn with Byzantine pottery, olive presses built of marble and cement, and broken cisterns, all these in the vicinity of Beersheba, a section which before 1900 was uninhabited except where the nomads encamped near the wells traditionally ascribed to Abraham. South of Beersheba were three larger centers whose ruins tell of settled and well-established populations, Sbeita, Khalasa and Kurnub.

The ruins of Khalasa are situated in an area of barren hills and soft limestone ridges overlaid with loose gravel and flint. Even though patch cultivation can be seen in isolated spots, the neighborhood as a whole is neglected and desolate. Terraces in the wadies tell of industrious husbandry in the past and are still dotted with shrubs. The ruins lie in a wide plain of light, good soil, some of which is now tilled by Bedouins, whose tents cluster near the well like great outspread bats' wings. Signs of former cultivation are seen in wadi after wadi, and everywhere there are fragments of pottery signifying Byzantine settlements.

On the road to Beersheba, we examined remarkable evidences of former cultivation on a broad plain that supported the ancient city of Sbeita. As far as one could see along the

[3] Major C. S. Jarvis, *Royal Central Asian Journal*, Vol. XXV, Apr., 1938, Part II, p. 208.

gentle slope black flint fragments were heaped together in mounds six to twelve feet across and two to four feet high. Captain Guy, the British archaeologist, told us that this area was once a vineyard. The vines had been planted and trained to run over the mounds where grapes were protected and sweetened by the heat of the stones. These stone heaps also provided for full absorption of rain and tended to prevent evaporation from the soil surface. Thus the roots of the vines benefited more from rainfall than they would have if planted in the usual manner.

Elsewhere, we found the slopes washed bare of soil, filling the wadi beds with sorted sands.

Leaving Beersheba for Kurnub further to the east, we passed for several miles through a well-cultivated region of good soil which was very dusty under the strong winds of a June morning. As our road followed down the slowly descending plain, the area of cultivation diminished and was more and more confined within the valley floors of wadies, where runoff from higher land has been diverted onto these restricted and well-placed fields.

As we rode toward the southeast, the country became more desolate looking. Now and then a Bedouin with a meager flock, or riding a camel, broke the monotony of the wilderness. Finally we dropped into the broad plain of a wadi draining toward the Dead Sea: a low range of hills with rock outcrop looked from a distance like a dam across the valley. We discerned a narrow gorge cut through this range. Here was the site of the ancient city of Kurnub, the causes for the decline and abandonment of which have aroused much speculation.

On a hill beside the gorge we found ruins of the walled city extending over about thirty acres. Kurnub was a flourishing Christian city before its abandonment and dates back to Nabatean times, covering a period of possibly a thousand years. A great church stood on the highest part of the city enclosure; it was built of fine well-cut stone that is now in ruins. We shall leave detailed description of the buildings to archaeologists and give our attention to the astonishing works of water conservation for irrigation.

One of the small tributaries from the ridge is the site of an ancient stone dam, long since washed out. Along the narrow valley walls above the dam there are remnants of a silt-filled reservoir. Evidently the dam became entirely filled with silt. Then during a flood the dam failed and succeeding storm floods swept out the fill and the dam though leaving sufficient evidence for reconstructing this story.

In the main gorge a little above the town a fine dam is still in place, built of well-cut marble stone. The dam is 38 feet high and 20 feet through at the crest and about 75 feet wide along it. The facing stone has been fluted by flood water bearing abrasive gravel and sand, which proclaims that many floods have gone over the dam since this city was abandoned. The valley has silted in to the height of the crest and broadened out into a considerable plain above the gorge.

Within the gorge above this dam are the buried walls of three other dams that were built later; they are respectively 100 feet, 175 feet and 375 feet above the first dam. Above the last-built dam are two piles of silt on the bank. The largest measures 100 feet wide, 130 feet long and about 10 feet deep at the thickest place. The smaller silt dump is 80 feet up and down the slope and 45 feet wide. These silt dumps are obviously cleanings from the reservoir impounded by the last dam. From this we may infer that this water development project was encumbered by silt; that extra dams were built to trap off the silt and when these failed, silt was cleaned out of the reservoir. These works are mute evidence of the inevitable damage of silt from an eroding drainage.

Still more remarkable were the terraces below the gorge, where the valley widens out leaving a gently sloping floor. On the west side of the channel a heavy stone wall was built roughly paralleling the stream channel. From this retaining wall a series of cross walls were built. The space between these cross walls of loose stone is estimated to have varied from 150 to 200 feet. The cross walls were level, but the crest of each one lower down the valley was lower than the one above. These walls were filled in with fine soil material into a remarkable series of flat terraces, descending like broad steps down the valley. The walls evidently were built to impound and settle out silt from silt-laden flood waters that were diverted out of the channel below the dam into a ditch emptying into the upper terrace. As this filled with silt into a terrace of fine alluvium, the muddy waters were diverted into the terrace next below until all were filled with silt.

Waters were diverted in the same manner to irrigate crops sown and planted on this fertile alluvium. In this way the inhabitants of Kurnub were fed with vegetables. For it is assumed that grains were grown in the wadi valleys more generally then than now.

It is indeed remarkable to find evidence of such a large center of population in a region that is credited by the rainfall map with only about three inches of rain. I hope the author of the map will pardon me for continuing to be skeptical until rainfall has been measured at Kurnub for a number of years. But whatever the rainfall was, not it but Kurnub's position on the caravan route from Aqaba accounted for the growth of so large a city on the margin of the Araba.

The Causes of Decline

In an earlier chapter dealing with the decline of Palestine we examined the various explanations of this phenomenon given by scholars, discussing particularly the theory which ascribes the transformation to a change of climate and, particularly, to decreased rainfall or pulsations of climate. It is this theory which is most frequently adduced as explanation of the desolate state of the Negeb today. Professor Ellsworth Huntington, for instance, maintains that the decay of the Negeb was due to climatic changes effected by cycles of rainy and dry periods. This is, however, refuted by a number of students, among them the late Sir Flinders Petrie, noted archaeologist and excavator in this region, Woolley and Lawrence; their conclusions are borne out by my own studies in the region. Woolley and Lawrence, as well as other geologists, agree that the climate has not altered to any great degree in the past two thousand years. They suggest that the agricultural development of this section in ancient times was the result of government activity.

Major C. S. Jarvis, who for years was Governor of Sinai, and who made one of the most profound studies of that region as well as of the adjoining Negeb, likewise ascribes the decay of the Negeb to political rather than climatic changes:

The most probable reason, I think, for the decline of this Roman civilization and cultivation was the fall of the Roman

Empire and the Arab invasion. When Rome went, public security
went. There were no police and no central authority, and these
towns out in the desert were cut off from civilization. They were
subjected to Bedouin raids. There was no initiative, no urge to
work, and general stagnation set in, as is always the case when
the Bedouin nomad gets the upper hand. They probably hung
on for one to three hundred years, and then I suppose gradually
the people died out or left these towns and went to some more
suitable spots. After this the Arabs' camels broke through the
fencing and started to eat the trees and the whole area went very
rapidly back to desert.[4]

The end of civilization in the Negeb came suddenly.
Shortly before the Arab invasion, the Byzantine government
had fearfully forbidden provincials to bear arms, while the
Emperor Heraclius drained away the garrisons. As the central power weakened, devastating incursions by fierce and
hardy Arab nomads from the desert increased, and the cities
of the Negeb melted away. Upland tent dwellers inherited
the thriving towns and farm works of Christian husbandmen:
the nomads had no interest in these refinements, which were
gradually destroyed or allowed to fall into ruins. The devastating sway of the nomad and his herds spread over the land;
trees were cut down, terrace walls crumbled, stone houses in
the towns were abandoned and only wells of the desert were
kept open to water the herds.

"The Mohammedan invasion in the seventh century swept
away the Byzantine civilization of Sinai as a locust swarm
devastates a corn field," states Jarvis in his book *Yesterday in
Sinai*. Even at best that highly developed civilization, based

[4] Major C. S. Jarvis, *Royal Central Asian Journal*, Vol. XXV, Apr., 1938,
Part II, p. 210.

on a complex division of labor, had been only an island oasis in a wilderness of nomadism. Despising the tilliŋg of the soil and hating trees, the nomads sought to live off herds and off the plunder of settled areas and the wealth of caravans in transit. This elemental antagonism between the tent dweller and house dweller has never been blotted out in the Holy Lands, despite the conquests of empire builders. Cyrus failed, as did Alexander. Rome, the greatest of all organizers, failed as did its Byzantine successor, and in the end all the Near East and the African portion of the Empire were overrun by the Arab nomads and the Empire fell and disappeared as completely as do the leaves of autumn, never to come to life again.

We do not know when the Bedouin completed his penetration of the Negeb. It appears to have taken rather long totally to destroy its civilization. Pottery is an indication of some settled population, and fragments found in Sbeita are of the twelfth century, five hundred years after the first Moslem conquest.[5] In limited areas the Bedouins' struggle for supremacy doubtless continued for many generations until settled civilization was finally completely effaced.

The Negeb today is still the land of the Bedouin. It is treeless, and houseless, but is marked here and there by somber black goat's-hair tents. Occasionally there is some patch cultivation where a few Bedouin families cut thin strands of barley with sickles. On the threshing floors camels trample out the grain in their slow and questioning way. Harvested grain is buried in a hole lined with straw and covered over with earth to form a small low mound, to which the Bedouin family will return for grain in the long fall and winter. An

[5] M. Avi-Yonah, *Palestine and Middle East Magazine*, Vol. 9, Sept., 1937.

unwritten law of the desert protects this unguarded storage of grain from plunder by other Bedouins.

The present Bedouin population of the Negeb is far below its population in the most prosperous period of Byzantine times. In the towns of Gaza, Rafa, Deer el Belah and Beersheba there is a settled population of 29,000. It is estimated that there are, besides, 21,000 Bedouins who wander about on three million acres, an area nearly equal to the remainder of Palestine.

According to census figures of the Palestine government the population of the Negeb has decreased since 1922. This is due mainly to increasing economic difficulties. Life is not what it used to be for the Bedouin, and it is steadily growing more difficult. As a result of motorized freight and transport, the breeding of camels, which once was a profitable source of income, has greatly declined. The price of wool has dropped, and opportunities to plunder caravan routes have practically disappeared. It is no longer possible to gather a large horde on racer camels and attack the despised village dwellers as in former times. Since the Mandate, this has been attempted only once. The attacking Bedouins were soon located by a scouting plane, and bombers were sent out with disastrous results for the would-be invaders and their camels.

From dire necessity Bedouins have had to resort to the despised art of farming, first in patch cultivation and later in more settled fashion. Others who are less resourceful camp outside towns and find simple work for wages. We were told that the Bedouin is very unhappy over this enforced change in his traditional mode of life. To him it is a real degradation. He is now halfway between nomadism and settled farming, having had to abandon the free and hardy life of the desert and compromise with the forces of modern times.

As a result of these changes, the area under cultivation in the Negeb has increased in recent years. By 1934-35 slightly more than 500,000 acres of the 3,000,000 acres were under occasional dry-farming cultivation with a crop in two or three years. According to field observations, patch or dry farming was observed well below Beersheba up to El Auja in a rainfall belt of only four inches of rainfall. It seems incredible with so little rain that even dry farming can be carried on with any chance of success. The only explanation is that the Bedouin in these marginal districts is gambling on the variability of rainfall; when heavy rains come he has a crop, but when they do not, he fails to harvest. During the season rain falls in spots throughout the area, so that somewhere grain will be harvested. Ignorant of even simple methods of conserving water, the Bedouin is a pitiful spectacle in our times, subject as he is to the vagaries of weather and unable to adjust himself to the new ways of making a living dictated by changes over which he has no control.

Unless the economies of grazing and farming become co-operating rather than antagonistic enterprises, there is no hope for lasting improvement in the present situation of the Bedouin. The co-operating and supplemental character of grazing and farming in the southwestern part of the United States gives us the clue. The vast grazing lands of the Bedouin must become the pasture lands for rich irrigated farming areas in alluvial valleys. By combining the two economies—grazing and farming—into an over-all supplementary economy, a solution can be worked out. Iraq, which has almost unlimited possibilities for irrigated farming, is ideal for growing crops to feed herds that are brought in off the steppe and desert during the dry season. Nomads may take up permanent residence in villages in the farming area, where children can be brought up in schools and the women saved the rigors of caravan travel. The male members of the family could follow the herds out into the desert during the rainy season when the desert becomes an ocean of green and luscious grass. Boys, as soon as they were old enough, could follow the herds into the grasslands with the older men. The same solution can be applied on a smaller scale to Palestine. We can envisage a plan whereby farming may be linked with grazing so as to keep alive the hardiness of the life of the open spaces, and at the same time remove the age-old antagonism between the tent dweller and the house dweller.

There has been much discussion as to future possibilities of the Negeb, though the casual observer may well conclude that its general appearance does not fortify hopes of much restoration. Experts, however, are more hopeful. Sir John Simpson, who was sent by the British government in 1930 to make a survey of the agricultural possibilities of Palestine reported: "Given the possibility of irrigation, there is practically an inexhaustible supply of cultivable land in the Beersheba area. Up to the present time, there has been no organized attempt to ascertain whether there is or is not an artesian supply of water." Lawrence and Woolley in their book *The Wilderness of Zin* express their firm belief that "today with ordinary methods of farming, the Negeb could be made to be as fertile as it ever was." Sir Flinders Petrie tells us of his own experiment when he was able to recover a cultivable area on a hilltop south of Gaza by the simple expedient of "plugging up the watercourses, and trimming the surface so that no excess of water could run off, and the result was that rich crops were raised."[6]

Water was the main problem in ancient times as it is today. But in this machine age we have more perfect instruments for our purposes. The southward slope of the Hebron Dome down into the Negeb suggests possibilities of artesian water sources. Under the proposed Jordan Valley Authority program, which would provide capital and authority for the complete restoration of the Negeb as well as other portions of Palestine, there is a distinct possibility of conducting irrigation waters to the Negeb from rivers in the north. During winter rains we saw these rivers, principally the Yarkon, dumping flood waters into the sea. These could be temporarily held back by dams and reservoirs higher up in the hills until the excess waters could be carried through conduits and dumped into great reservoirs or artificial lakes in the Negeb to be used for irrigation. At San Dimas in California this has been done successfully; indeed, when one has seen the great conduits pouring water from the Colorado River into Southern California after crossing deserts and boring through mountains for 250 miles, one is certain that similar engineering projects on a smaller scale are possible in Palestine. The Negeb, in general, is not unlike some of our southwestern deserts which have been turned into productive gardens by irrigation waters.

With the cheap power of the JVA, industries likewise could be established in this southland, quite as limited industry in weaving and dyeing helped to support ancient peoples there. I believe that as soon as even partial irrigation is provided for those areas in the Negeb suitable for irrigation, the region

[6] *The Revival of Palestine.* Jerusalem, 1937.

can be made to maintain a far larger population than it fed in ancient times. With thousands of land-starved Jews clamoring to get into Palestine, this vast area, almost the size of the present populated portion of Palestine, should be widely opened to resettlement and reclamation.

We were strengthened in these convictions by soil samples taken in various sections of the Negeb, and by the special studies we made. In the Coastal Plain there is a light sandy loam on which considerable cultivation is possible even without irrigation. During winter and spring, watermelons and tomatoes are grown there in quantity. Olives and apricots do well, but citrus is not so successful as further north. Water can be found at depths of from six to a hundred feet, but is sometimes brackish. From Gaza to Beersheba we passed over an undulating rising plain of deep, good soil, shown in road cuts to be a layer of loess over weathered country rock. From Hebron to Beersheba we found the hilly country with outcropping limestone, sloping away until it flattened into an undulating plain in which the drainages are filled with deep, light and sandy alluvium, but intervening hills are bared of soil. Between Beersheba and Kurnub a wadi flood has exposed chalky limestone overlaid by a mantle of loess twenty-five feet thick. Loess soils are fine textured and when irrigated they are exceedingly productive. Alkali content is low, for the ground has been leached out by winter rains.

In my opinion, there is promise in the Negeb for much improvement in certain types of cultivation, even without irrigation, providing such cultivation is done according to methods worked out during recent years by the United States Soil Conservation Service. Within the dry-farming belt of eight inches and more of winter rainfall, it is most important to conserve all rain that falls on farm fields. The productivity of the region could also be greatly improved by contour farming; basin listing in the plowing of fields; building broad base level terraces; leaving the crop litter at the soil surface; following certain improved methods of fallowing; and similar devices.

These scientific methods have been adopted in our own West for dry-farming conservation of rain waters in semi-arid regions; none of them is being practiced in the Negeb. In our survey of the region, we could see that all the drainage channels of the slopes are cutting deeper: it is evident that storm waters are increasingly eroding the country. In short, soils lose in storm runoff much of the beneficent rain that falls. The land is literally being "desiccated" by this loss.

A potential economic asset in the southeastern part of the Negeb is seldom mentioned. While we were at Aqaba, Solomon's old seaport on the Red Sea where Dr. Nelson Glueck was then excavating, our attention was drawn to the fact that in those waters there are enormous quantities of fish. The natives fish for their own use, but there has been no attempt to exploit this untouched resource for export. Palestine's coast supplies no great quantities of fish, yet here the water teems with gray mullet which may be pickled or salted, with a species of herring, red mullet, various types of mackerel, crayfish and other varieties. As we swam about, highly colored tropical fish swam with us in the clear warm waters. With the introduction of modern methods of dry-ice refrigeration and the building of improved auto roads into Palestine, it should be possible to arrange for large-scale commercial exploitation of this natural resource.

In any consideration of the possibility of reclaiming the

Negeb, it is important to recall the experience of Jarvis in opening up old Roman settlements in the Wadi Gedirat just over the border in Sinai. While hunting partridge on a hillside, Jarvis picked up the position of an old Roman dam and reservoir and the lines of channels. He constructed another dam to carry off all the water to the Roman reservoir as in ancient times, but the reservoir had first to be excavated, for it was filled to the brim with silt. From it one was able to irrigate all the land around and bring it under cultivation. Jarvis wrote:

There is some Australian wheat grown in that valley. The people who say there is no humus or anything else left in the soil ought to see it. In the valley we discovered that the olive grew exceptionally well. I had sent to me by the Ministry of Agriculture in Egypt olives from every country—from Tunis, Cyprus, Spain, Italy, Palestine—and they all did equally well. They grew to an enormous size and very quickly, and were all bearing within four or five years. When I left there were over four thousand trees coming into bearing and new trees were being added every year. The apricots also grew extremely well, and these can be dried and are easily transported. The vines were a very great success. The orange and the grapefruit were not very good. That is because that high desert lacks the humidity which is necessary for them. There were also other fruits, such as figs and the date palm. Asparagus grows exceptionally well there. There is a very big future for asparagus there, because people in Jerusalem, Cairo, and Alexandria consume great quantities of it. . . .

One of the criticisms I have heard is that this scheme of reopening up the old Roman cultivation system may be all right for the Bedouin, because his standard of living is low and he can get enough out of it to just exist, but that it will not do for Europeans. There may possibly be something in this, and the new settler will

certainly have to work very hard and make the most of everything, as his standard is higher than that of the Bedouins. On the
other hand, you must remember that a settler of another race
would make a better show of it than the Bedouin, because the
Bedouin does not take kindly to cultivation. Even when they get
the water flowing to the land, they only cultivate the easiest and
flattest bits and make no attempt to level it or to clear away stones
or scrub.[7]

Jarvis goes on to say: "I am afraid if you took your hand
away . . . the Bedouin will let things go; the dam will break,
the channels silt up, the animals will break in again, and the
gardens will go back to the desert as it was before."

The conclusion is obvious. Only an energetic people, using
scientific methods of dry farming and constructing dams to
conserve flood and rain waters, can make the Negeb fertile.
The Jordan Valley Authority, which would have the right
and capital to bring to the Negeb surplus waters from rivers
at the north, could particularly facilitate the upbuilding and
repopulation of the Negeb.

[7] Major C. S. Jarvis, "Southern Palestine and Its Possibilities," *Royal Central Asian Journal*, Vol. XXV, Apr., 1938.

CHAPTER XIII

TRANS-JORDAN—PAST, PRESENT AND FUTURE

. . . .

THE frontiers of Trans-Jordan run along the Yarmuk River in the north, and through the Arabian Desert in the east; they touch the Gulf of Aqaba on the south, and follow the Araba, the Dead Sea and the Jordan River on the west. Trans-Jordan is ruled by an Arab prince, Emir Abdullah, under the close supervision of a British adviser subordinated to the High Commissioner for Palestine.

Trans-Jordan was once one of the granaries of Rome and supported a highly developed and advanced population of about a million people—three times its present primitive and backward population. The archaeological investigations of Glueck, Iliffe and Harper have revealed the ruins of flourishing cities, villages and farms. My own examination of ruins at Petra, Ma'an, Wadi Hesa, Amman, Mafrak, Um Jemal, Jerash and elsewhere convinced me that in ancient times the country enjoyed a Golden Age of agriculture. I am fully convinced that if the "Jordan Valley Authority" reclamation plan is put into operation, the neglected land of Trans-Jordan may become even more prosperous than in olden days.

For a soil conservationist a study of Trans-Jordan today presents a tragic comparison with the past. The region's agricultural and cultural development reached its greatest height under the Nabateans, who were its undisputed masters from the fourth century B.C. to the second century A.D., were thereafter dominated by the Roman and Byzantine empires, and were finally blotted out by the Arab invasions beginning in the seventh century A.D. Petra, the magnificent Nabatean capital, carved out of red sandstone cliffs reminding us of those in Zion National Park in Utah, crumbled away in the fastness of its desert gorge. It remained unknown to the modern world until its spectacular rock-carved tombs and buildings were rediscovered in 1812 by the Swiss explorer, John Lewis Burckhardt.[1]

In studying Trans-Jordan we had the valuable guidance of Captain P. L. O. Guy, Director of the British School of Archaeology, and of Dr. Nelson Glueck, Director of the American School of Oriental Research at Jerusalem. Dr. Glueck has located and mapped the sites of more than 750 towns and villages, marked by fragments of carved stones and pottery, ruined walls of houses, disarranged stones of terraces, cisterns and dams.

These ancient sites are dated by sherds of pottery distinctive of the ages of occupation, beginning with the late Stone Age, 5000 B.C., when man first became a food producer or farmer, and continuing through the Bronze and Iron ages, to the Hellenistic period from 300 B.C to 100 B.C., the Roman period from 100 B.C. to 400 A.D. and the Byzantine period from 400 A.D. to the Arab invasion of the seventh century. After that period nomads out of the desert overran the country and supplanted permanent settlers with their nomadic tent-dwelling culture. They used goatskins for containers instead of pottery, and lived upon their grazing animals rather

[1] J. L. Burckhardt, *Travels in Syria and the Holy Land.* London, 1822.

than on cultivated crops. They left no permanent remains. It was not till the last century that permanent settlements of any sort were established in Trans-Jordan, they were villages of Circassian colonists who fled from Russian rule in the Caucasus and were encouraged by the Turkish government to act as buffers against invasions from the desert.

The fortunes of the people of Trans-Jordan throughout the past three thousand years have been closely associated with those of Palestine, because of the physiographic unity of the two areas both draining into the Jordan Valley. Moreover, the peoples of these two areas sprang from the same Semitic racial stock which from time to time swept out of Arabia toward the better watered coastal highlands and valleys. The earliest of these waves was that of the Jews—about three thousand years ago, when they abandoned the nomadism to which they had reverted after their flight from Egypt. In Solomon's time their descendants dominated a larger area than the combined territories of Trans-Jordan and Palestine today. More than two thousand years ago another wave of Semitic nomads known as the Nabateans settled in the southern half of Trans-Jordan, established their capital at Petra and developed a remarkable agriculture. A little later, in what is now northern Trans-Jordan and southern Syria, the Greeks built up the famous Decapolis or League of Ten Cities, which served as a barrier to the northward extension of Nabatean rule. The southernmost city of the Decapolis was Philadelphia, which is now called Amman and serves as the capital of Trans-Jordan.

Trans-Jordan's ancient civilization was built on farming and trade, for Trans-Jordan was the back corridor through which north-south caravans passed from Anatolia and Damascus in Syria to Egypt, and from Arabia, via Petra, to the Mediterranean Sea at Gaza. Petra was the crossroads of these two great caravan routes and owed its wealth and importance to that fact. After the break-up of Roman and Byzantine power in the seventh century, Trans-Jordan was overrun by one conqueror after another and exposed to wave after wave of plundering desert hordes who overran the well-tilled and laboriously protected fields. As terrace walls were neglected and fell into ruin and works for the conservation of rainfall and irrigation likewise fell into decay, erosion, the great enemy of civilization, took over the country.

The tragic decline of Trans-Jordan was epitomized for us in the black goat's-hair tents and the nomad family of fierce men, painted women and unkempt children we found amidst the fragments of columns and carved capitals of a magnificent temple or Christian church at Jerash. Two women were carrying water in slimy goatskin water bags, and one bore a heavy load of fuel on her back. Near the black tents, which resemble huge bats with wings outspread, there were a few patches of barley, representing a very primitive sort of agriculture. Everywhere water conduits are out of use from neglect; most cisterns are filled with dry debris, and fields are abandoned among ruins of former terraces. Here and there one finds remnants of former forests and long-unused olive and wine presses constructed of stone.

The official report on the administration of Palestine and Trans-Jordan submitted to the League of Nations in 1924 stated that "the population of Trans-Jordan is thought to be in the neighborhood of 200,000, of whom some 10,000 are Circassians and 15,000 Christians; the remainder are in the main Moslem Arabs." A report for 1935 gives the figure for the entire population as between 300,000 and 320,000, for the Circassians 7000 and for the Christians 20,000. According to available statistics the birth rate among the Arabs of Trans-Jordan is much lower than that in the mandated area of Palestine. Moreover, evidence indicates that Trans-Jordan has lost some population by emigration to Palestine. Since no official census has been carried out in Trans-Jordan, we shall accept the higher estimate of its population as most nearly correct. Even this figure—320,000—is only two-fifths of the most conservative estimate of its population under Roman rule.

While the density of population in western Palestine is about 150 per square mile, there are only nine to ten inhabitants per square mile in Trans-Jordan. In other words, mandated Palestine where Jewish settlement has been taking place is about fifteen times as densely populated as Trans-Jordan. But we must not forget that the percentage of desert and semi-arid lands in Trans-Jordan is much higher than in western Palestine: the semi-arid Negeb forms only 45 per cent of the total area of mandated Palestine, but deserts and semi-arid districts form 80 per cent of Trans-Jordan. In order to arrive at a fair comparison, we should disregard the arid and semi-arid districts in both Palestine and Trans-Jordan and consider only the districts mainly settled by a permanent farming population. Interestingly enough, the area of permanent settlement in Trans-Jordan is officially estimated as 17,500,000 dunams or 25 per cent more than the corresponding 14,000,000 dunams of western Palestine. The permanently settled population of Trans-Jordan is one-seventh as large as that of Palestine.

The physical features of Trans-Jordan have been described in a general way in the chapter on the geography of Palestine, but certain additional details should be noted in order to facilitate understanding of the country's future possibilities.

From west to east, the land falls into four natural divisions: (1) the Ghor, or Jordan Valley Depression, (2) the hill lands of the escarpment, (3) the plateau west of the Hedjaz Railroad, and (4) the dry lands between the Hedjaz Railroad and the desert of Arabia.

The Jordan Valley, or Ghor

Within Trans-Jordan the Jordan Valley extends from the Yarmuk River, which opens out into the valley just below the Sea of Galilee, to the Dead Sea—an area which has an average width of about two miles, expanding in certain districts to nine miles. The fellaheen of the Jordan Valley irrigate their lands, except for a restricted area in the north which receives enough rain for farming without irrigation. Wheat and barley are sown in the winter and white durra in the summer in rotation. Some vegetables are grown, and in favored sites, such as Beisan and Jericho, bananas, citrus and pomegranate orchards are found. Ancient methods of sowing, plowing and irrigation are used, and most of the water from perennial streams flows into the Jordan without being utilized for crops.

Vegetable growers are harassed by plagues of insects, plant diseases and destructive winds. In the absence of modern measures for the control of insect infestations and plant diseases, this type of farming has thus far not proved profitable. Citrus and banana plantations, the crops of which are sold to Palestinians, are more successful.

Most of the land of the Jordan Valley is owned by absentee landlords. The present population is small, and chiefly confined to the northern area. Suitability of the land for irrigation varies: the prospects are good in the north, where about half the area may be irrigated from the Yarmuk and other rivers draining the high limestone plateau. This region is a natural counterpart of the irrigable Beisan Plain.

The Hill Region

The eastern escarpment of the Jordan Valley, rising to the indented rim of the Ghor, reaches altitudes of more than five thousand feet. It varies in width from three to fifteen miles, and has an area of about 1,500,000 acres. In ancient times this high mountain mass condensed rainfall sufficient to grow forests and to support extensive cultivation and dry farming. It does the same today wherever soil has survived the wastage of erosion. The escarpment itself consists of rough hilly land where out of a total of 1,200,000 acres not more than 500,000 are suitable for cultivation. The rough slopes, especially in the lower and middle zones, are best suited to grazing, but the upper slopes produce good forests. At present, about 100,000 acres are in forest growth. In our land-use survey we examined stumps of trees three feet and more in diameter in the highlands between the Wadi Hesa and Petra; here in ancient times a southern forest of considerable extent doubtless supplied fuel for the copper furnaces of Ezion Geber which we saw Dr. Glueck excavating in 1939. The forested lands are now generally grazed without protection of trees and young growth so that restocking is poor or does not exist at all.

The farm lands of the hilly section include flat lands along narrow stream valleys and lands on hill slopes. Upper slopes are planted to vegetables and barley and lower slopes are

[207]

covered with vineyards and dates. When irrigated under the plan of the Jordan Valley Authority, these lands will furnish excellent sites for terrace orchards and vineyards to rival those along the Rhine in Germany.

The districts of Es-Salt and Amman have excellent agricultural possibilities, as the splendid ruins found there indicate. The rainfall is sufficient to support grain farming, tree crops and vineyards. East and south toward Madaba, ruins are not so numerous, showing less productivity of land in that region, but here, too, about 300,000 acres of land are suitable for extensive dry farming.

The northern area is agriculturally the best land in Trans-Jordan and now supports the largest population. Much of the land of the Yarmuk Valley is under irrigation from crudely designed intakes and canals. According to Ionides, a much larger area here may be brought under irrigation.[2] The waters of the Yarmuk are sweet and once served to water lands in the Jordan Valley, as ruins of inverted siphons used during Roman times indicate.

The Plateau

The third division of the lands of Trans-Jordan includes the plateau between the picturesque, indented high rim on the west and the Hedjaz Railway on the east. From the rim that rises to altitudes of more than five thousand feet, the plateau slopes down toward the east. These lands are generally flat, and include the shallow valleys of the Hauran, the Beni-Hasan lands, the Ajlun district, the flat lands of the Valley of Balka, the fertile plains of Kerak and Tafileh, the plains of Zara, which at one time supported a large farming and live-

[2] M. Ionides, *Report on the Water Resources of Trans-Jordan.* 1938.

stock-breeding population, and finally the Esh-Sherah Valley in the district of Ma'an. The area of these flat lands is about 1,307,500 acres of which 525,000 acres are said by officials of the Trans-Jordan government to be suitable for farming. These estimates need checking by a land-use survey which should classify all the lands of this region by kind and area. It is my opinion that a balanced use of these lands for grazing and farming will support a greater population than a sharp separation of intensive farming from nomadism.

Most of the plateau is sparsely populated, except for a few administrative centers chief of which is Amman. Elsewhere inhabitants follow a life of semi-nomadism over areas that once supported fair cities and exported grain to Rome. It is in this region that there once flourished the cities of Philadelphia, Gerasa, Pella and Gadara in the north, and Petra, Kerak, Ma'an and many villages in the south.

Winter crops now grown by fellaheen on these lands are wheat, caraway seed, oats, linseed, peas, beans and durra. Summer crops are greens, watermelons, marrows, tomatoes and onions. Potatoes have not proved to be a satisfactory crop thus far. Camel breeding is one of the chief enterprises of the Bedouin in the region and cattle are raised by the semi-nomads.

The Arid Area

The fourth division of Trans-Jordan includes the lands east of the Hedjaz Railway to the desert border with Arabia. The small area under cultivation is in the Balka district in the south. There is insufficient rainfall for farming in this district and it is best suited for grazing. However, the existence of a number of springs such as El-Azrak, Al-Umari, Al-Khaza, Al-

Hafeir and Bayir suggests that it should be possible to bore wells. Elsewhere the land is natural grazing country that reminded me of eastern Colorado. Bored wells would be a great boon to the nomads in watering their livestock. There is an excellent opportunity here to combine farming with the raising of livestock as in the western United States.

Climate

The climate of Trans-Jordan, like that of most of the Near East, is essentially sub-tropical. It is typically Mediterranean in distribution of seasons, having a rainy winter, lasting from October to May, and a long dry summer season from May to September. The highlands of Trans-Jordan, unlike the coastal areas, have comparatively cold winters with sporadic snows, while the summers are comfortably cool, except when the khamsin, or hot east wind, blows out of the Arabian desert. Because Palestine and Trans-Jordan lie on the northern margin of trade winds that blow over dry continental areas, they have a climate of great variability similar to that of the southwestern plains in the United States. Variability in rainfall is characteristic of both Trans-Jordan and Palestine.

Reliance of farming and grazing on regularity of rainfall has often brought disaster to man and beast alike. In a region of high variability in rainfall, agriculture must be adapted to such variations as exist in the western plains of our country.

The weather in Trans-Jordan, while we were there, was delightful and invigorating. George Adam Smith, whose *Historical Geography of the Holy Land* is an incomparable source of information on Palestine past and present, describes the climate of Trans-Jordan in the following glowing lines:

TRANS-JORDAN—PAST, PRESENT AND FUTURE

We traversed Eastern Palestine during twenty-two days of mid-summer, and were therefore able to test the climate. We had thrice dense mists, and several very cold evenings. Every morning about ten a breeze sprang up from the west, and lasted until sun-down, so that although the noon temperature in the Jordan Valley, as often as we entered it, was at least 103 degrees, on the table-land above we seldom had it over 90 degrees. Whether upon the shadeless plain of Hauran, where the ripe corn swayed like the sea before the wind, or upon the ridges of Gilead, where the oak branches rustled and their shadows swung to and fro over the cool paths, most of the twelve hours were almost as bracing as the dawn, and night fell, not, as in other parts of Palestine, to repair, but to confirm, the influence of the day. Eastern Palestine is a land of health. . . .

In summer, nights are always cool. Days are usually sunny, except during the rainy season from November to March, but long stretches of sunny days occur even between rains. In the winter, moisture from the Mediterranean is blown by stiff winds against the high plateau, condensing a generous rainfall. Health resorts seem to have been located here in Roman times. Recently some sulphur seepages were cleared of debris and revealed the masonry of Roman baths.

There is much more water in Trans-Jordan than I expected to find. There are many flowing springs in spite of the loss of soil and the lack of vegetation on the denuded rocky slopes. The rock foundation of limestone, unlike the granite and gneiss of Southern California, favors the absorption of winter rains. The limestone is shot through with solution crevices which provide vertical drainage channels for rain waters that would otherwise rush off the slopes. The ruins one finds near

[211]

springs indicate that no great change in water supply has occurred since ancient times.

Many springs, however, have been befouled and trampled by Bedouin herds so that their flow has been choked and their efficiency impaired. A few years ago a serious epidemic of typhoid at Amman, the capital of Trans-Jordan, was called to the attention of the British archaeologist, Captain P. L. O. Guy. He investigated the befouled spring which supplied the city with water and suggested that it be cleaned out. On digging back into the hill, the workmen found fine cut stone masonry lining the spring and the intake of an aqueduct which had supplied the city during the Roman period. The flow of the spring was greatly increased after the cleaning and the city has since been supplied with a larger supply of pure water. Many instances of the rejuvenation of springs by cleaning may be cited for both Palestine and Trans-Jordan.

Even in the eastern region of Trans-Jordan on the border of the desert, traveling across the great Wadi el Hesa we came upon an area of greenery in striking contrast to the barren rocky surroundings. Ruins of terraces from crest to valley floor indicated that the entire slope had once been terraced and cultivated. Remains of ancient rock conduits that had carried spring water out to the terraces were strewn about the area. A good stream flowed in the wadi; sand and debris from flood waters had piled upon what otherwise might have been garden lands. Numerous olive and wine presses made of stone have been discovered in this location by Glueck, but today spring waters flow unused down the slopes to form a large swampy area grown with weeds and rushes. Only a small portion of the water is used and many of the ancient gardens are now thorny wastes.

Even British officials who generally tend to minimize the potentialities of Trans-Jordan concede the existence of large, unused water resources. F. A. Stockdale, Agricultural Adviser to the Secretary of State for the Colonies, who studied Trans-Jordan in 1935, reported as follows:

In the valleys, utilization is made of water for irrigation wherever it is available. This water is derived from springs which form the source of the perennial streams. Long irrigation channels following the contours are frequently to be seen in various parts of the country. I was able to inspect one of these irrigation schemes on the way up from the Allenby Bridge to Amman, and others in the north of the country. There is considerable wastage of water from the channels, and it is quite clear that at least a 50 per cent increase in agricultural production would be possible if the irrigation water were under controlled distribution and if efficient canalization were provided and adequately maintained. There are definite possibilities of development if the existing irrigation schemes are improved and others established.[3]

Mr. Ionides, who made a study for the Trans-Jordan government of the water resources of the country, estimates the rainfall supply and disposition of water in its western portion as follows:

Rainfall over the entire area	3,524,000	acre feet
Ground water flow, of the Jordan above the Yarmuk	222,000	" "
Direct runoff, of the Jordan above the Yarmuk	235,600	" "

According to Ionides the amount of water in the western part of Trans-Jordan, directly available for artificial irrigation, is:

[3] Report to the Secretary of State, p. 274.

Yarmuk—ground water flow	200,000 acre feet
Yarmuk—direct runoff	178,000 " "
Jordan flow at outlet from Lake Tiberias	425,500 " "
Total flow of the Jordan and tributaries	803,500 acre feet

This flow, if effectively applied to the land, would irrigate no fewer than 200,000 acres of land, when the quantity of water used is placed at four acre feet per annum for the hot Jordan Valley.

After studying these figures, I am inclined to agree with the Palestinian experts who consider them an underestimate. My conclusion is based on the calculations of the evaporation from the water surface of the Dead Sea: there is in all probability more water available from the Jordan for irrigation than Ionides has indicated.

The Jordan Valley and the western hills are well watered by springs and streams, but the upland plateau depends for its water supply on rain-water cisterns. Irrigation water must be stored behind dams and in reservoirs where suitable sites can be found. There are a number of wells east of the Hedjaz Railway, the most important being at Bayir and Al-Hafeir. Boring experiments carried out in this region have shown that its water resources are considerable. If these were developed, livestock raising would be more secure.

Further study of the geological structure of Trans-Jordan and its land and water resources may disclose possibilities for storing water in underground caverns. If outlets were stopped by dams and interior walls were sealed by cement guns, large volumes of water which flow in the winter might well be stored for summer use. This method of storage has the advantage of low evaporation which may reach the depth of as much as six to ten feet. Subterranean storage reservoirs in limestone caverns would save much of this loss.

Trans-Jordan may be divided into two rainfall areas; one too arid for dependable dry farming, and the other favored with enough rainfall to make dry farming profitable. According to Ionides, the line of division falls on the 8-inch line of equal annual rainfall. Scientific farming with winter grains under winter rainfall is successful in central Washington under rainfall as low as 6.72 inches per annum. To place the limit of dry farming at 8 inches of rainfall in Trans-Jordan at 3000 to 5000 feet elevations of the plateau is considered con-servative. Under especially favorable conditions, dry farming may be pushed even further into the arid zone.

The eastern part of Trans-Jordan, beyond the Hedjaz Railway, is in the area of 8 inches or less of annual rainfall. It is well suited for grazing under a combined economy where the growing of seasonal fodder and grazing support each other.

Present Condition

In spite of Trans-Jordan's large stretches of fertile soil, comparative abundance of water and extraordinarily health-ful climate, its population is in an abject state of poverty. Most cultivation is very primitive, and crop failures period-ically result in famine conditions. The Trans-Jordan govern-ment, burdened by deficits, is unable to undertake adequate relief and reform measures. In marked contrast, the Palestine government has been able to give substantial subsidies for food and seed to fellaheen in districts afflicted by drought: the prosperous state of the Palestine government's treasury, as we have pointed out above, is due to the revenue resulting from Jewish immigration and enterprise. In 1933 when Trans-

Jordan was afflicted by a severe drought, a considerable number of its inhabitants made their way to Palestine where they found work to sustain them through this difficult time. It is said that the ruler of Trans-Jordan, Emir Abdullah, and some of his more far-sighted advisers are seriously alarmed by the possibility of a growing depopulation of Trans-Jordan: there are authentic reports that they at one time thought of improving the economic situation in the country by opening it to restricted Jewish settlement.

The backward state of Trans-Jordan as contrasted with Palestine is evident in a comparison of the consumption of foreign goods in both countries. The figures for imports in the accompanying table are given in Palestinian pounds, the official rate of exchange of which is slightly under five American dollars:

Year	Trans-Jordan	Average per Person	Palestine	Average per Person
1935	607,624	2.03	17,853,000	14.92
1936	794,956	2.56	13,979,000	11.01
1937	934,211	2.85	15,904,000	12.06

The decline in Palestine imports from 1935 to 1937 reflects the influence of the protracted Arab riots which started in April, 1936, and brought about a sharp decrease in immigration and settlement. But even at the height of the riots in 1936-37 the average inhabitant of Palestine consumed about five times as much foreign goods as the average resident of Trans-Jordan.

In 1936-37, the last year when statistics on both countries were published in an official report to the League of Nations, there were only 7105 pupils in all the government schools of

[216]

Trans-Jordan, and 5526 in private and missionary schools. These schools, many of which are on an extremely primitive level, are frequented by only 14 per cent of the children of the country. The others receive no schooling at all. In the elementary schools of Palestine 145,420 children are enrolled. If we exclude the Jewish children, 95 per cent of whom attend school in spite of the lack of a compulsory education law, we find that the percentage of Arab children attending school in Palestine is three times as large as in Trans-Jordan.

In 1936-37 the whole of Trans-Jordan possessed only 4 missionary hospitals with 107 beds and 3 government hospitals with 38 beds, while Palestine had 40 hospitals and 2983 beds. There is no insane asylum in Trans-Jordan and when the insane become violent, they are put into prison for safety's sake. The rate of infant mortality in Trans-Jordan in 1937 was 203 as compared to 152 for Palestine as a whole and 179 for Palestine's Moslem population.

Though Trans-Jordan was formally separated from Palestine in 1921, it is still treated as part of Palestine in many respects. Its railroad is considered part of the Palestinian system; its border defense is subsidized by the Palestine Treasury. Indeed it is difficult to find any valid reason why the backward and semi-starving country of Trans-Jordan should be excluded from the benefits which may accrue to Palestine from the great reclamation project of the Jordan Valley Authority. The JVA could provide the inhabitants of Trans-Jordan with water for irrigation, with better varieties of livestock, improved seed selection and insecticides, and teach them modern methods of farming and grazing as well as enable them to enjoy the benefits of cheap electric power. Moreover, the Jordan Valley Authority as a whole

[217]

would lose much of its effectiveness if only the region west of the Jordan were included in the sphere of its activities.

During many centuries Trans-Jordan was considered an integral part of Palestine not only for historic reasons, but essentially because the two form a natural economic and geographic unit. There is no reason to end this close relation now when all the inhabitants of Greater Palestine—Jews and Arabs alike—may benefit greatly by a reclamation program affecting their common welfare.

CHAPTER XIV

PALESTINE'S ABSORPTIVE CAPACITY

. . . .

THIS book has outlined a program intended to increase the absorptive capacity of Palestine to a very marked extent. Absorptive capacity, it must be remembered, cannot be measured by a yardstick. It expands or contracts in accordance with the degree of justice and security provided by the government of a country and in accordance with the genius of the people who occupy the land. We have seen how the absorptive capacity of ancient Palestine was built up by the labor and ingenuity of countless toilers during Greek, Roman and Byzantine times, until the resources of the country provided sustenance and prosperity for many times the population which inhabited Palestine at the beginning of the twentieth century. But we have also seen how the Arab invasions abruptly put an end to this prosperity, and how in the ensuing centuries exploitation, plunder and neglect of ancient conservation devastated and depopulated the land. Its absorptive capacity had, in other words, been sharply contracted.

Recent Jewish colonization, though confined thus far to only 6 per cent of the land, has brought about expanding absorptive capacity once again. Hence the population of Palestine has grown from below half a million in 1900 to

[219]

more than a million and a half in 1942! The consecrated genius and vision of the Jews in draining swamps and turning sand dunes into orchards and poultry farms, in planting millions of trees on the rocky hills, in building terraces, digging wells, developing irrigation, establishing numerous and varied industries and founding hospitals and clinics, has brought a greatly increased measure of prosperity to Palestine while making possible not only the settlement of almost half a million Jews in the last twenty-five years but the doubling of the Arab population in the same period. The desire to preserve their race, to prevent extermination through ghastly persecutions, has provided a powerful driving force for the Jews in Palestine, and they have concentrated the abilities of their most highly skilled engineers and technicians upon the problem of providing food and labor for a maximum population. It is practically impossible to estimate what the final absorptive capacity of Greater Palestine could be if all its unoccupied or under-populated areas were rejuvenated by the same vigor and understanding love of the land as have characterized Jewish efforts on a tiny fraction of the land, and if such an all-inclusive reclamation program as that of the JVA were put into effect.

The decline of the Near East and particularly of Palestine was due primarily to the fact that for centuries the population was not protected by government from the raids of marauders or from exploitation by usurers and rapacious officials. Whenever the country enjoyed a reasonable degree of security and justice even for a limited time, the general well-being rose rapidly. Thus under Greek, Roman and Byzantine rule agriculture developed astonishingly, and even in the Moslem period there were a few benevolent or far-sighted rulers in whose time the resources of the country quickly responded to the efforts of the tillers of the soil.

The second important factor determining the absorptive capacity of a country is the genius of the people inhabiting it. Only a people with a true love of the land and a goal toward the achievement of which it consecrates all its efforts, can bring a country to a maximum state of development. The Dutch in their little homeland are an admirable example: on lowlands continually encroached upon by the North Sea they have industriously developed a homeland for about eight million people. When their growing population required more land, they increased the absorptive capacity of their country by dyking off sections of the North Sea, draining them, leaching out the salt and preparing the soil for crops. As we have already noted, they changed sections of the ocean floor into an agricultural paradise yielding bountiful crops.

The genius of the Jewish settlers in Palestine has similarly transformed "bad lands." They have fled from the unspeakable persecutions of anti-Semitic regimes in Europe. They have a goal to which all of them are consecrated and they are undaunted by hardship, disease, crop failures or terrorist attacks. By training thousands of young men and women in technical and practical tasks and organizing them into working groups which can wrest their food, clothing and shelter from the soil and find satisfaction in so doing, the Jewish pioneers in Palestine have developed something new under the sun. A foundation has been laid and a method evolved for successfully adapting a people to its land resources.

Given security and justice, and a people with spirit and will, economic achievements in any given land are limited by physical environment, land resources, topography, climate and the opportunities for industry and commerce.

Let us first consider the land resources of Palestine. Its total area under the Mandate (excluding the surface of the lakes) is 26,319,000 metric dunams equivalent to 6,579,750 acres. This tract of land, about the size of Vermont, consists of a maritime plain, highlands, narrow valley and the arid district of the Negeb. Unfortunately, no soil survey has been made of the entire area. Dr. Strahorn mapped the soils of the lowlands, the maritime plain, the Emek and the Jordan Valley west of the Jordan River, but not the highlands which comprise five-sixths of the total area. He concluded that the 4,874,952 dunams he had mapped, out of the total 26,319,000 dunams, fell into the following categories:

Irrigable farmlands	2,964,320 dunams
Non-irrigable farmlands	1,418,648 dunams
Wasteland	490,064 dunams

Julius Fohs, a noted American expert, who conducted extensive hydrographic surveys in Palestine, estimates the irrigable area of Palestine at a somewhat higher figure. He states that after providing water for the civil and industrial needs of a population of 2,500,000, it should be possible to irrigate 3,500,000 dunams in Palestine exclusive of the Negeb. Mr. Fohs envisages the possibility of irrigation projects in the hill country of Judaea and Samaria, which was outside the scope of Dr. Strahorn's investigation.

Since both Dr. Strahorn and Mr. Fohs conducted their surveys years before the project of the Jordan Valley Authority was suggested, their estimates do not include the additional irrigation waters the JVA will make available for the Plain of Esdraelon, as well as for other parts of Palestine which are not now considered irrigable. Yet even the comparatively low figure arrived at by Dr. Strahorn after his partial survey of the country is more than six times as high as the number of dunams now actually under irrigation (480,000 in all of Palestine). If the entire area considered irrigable by Dr. Strahorn were irrigated, the food production of the country would increase greatly.

The difference in productivity between irrigated and non-irrigated farmland is more striking in Palestine than in most countries. The farmer cultivating 200 dunams (50 acres) of non-irrigated land is dependent on the variable climate and rainfall, and frequently suffers total failure of crops, while a farmer intensively cultivating 20 to 30 dunams (5 to 7 acres) of irrigated land makes a modest but secure living.

On the question of the total cultivable area of Palestine, experts' opinions vary sharply. In a report issued in 1930 the Commissioner of Lands estimated the cultivable area of Palestine as 12,233,000 metric dunams. His definition of "cultivable" is "land that can be brought under cultivation by the labor and financial means of the average Palestinian Fellah." This naturally excludes the marshes, the coastal sand dunes, the wilderness of Judaea and the semi-desert areas south of Beersheba. A few years later the government revised this estimate downwards. The Director of Surveys, basing his conclusions to a large extent on aerial surveys of typical districts in the hilly country, estimated the immediately cultivable area as only 8,044,000 dunams. In the reports submitted to the Royal Commission of 1937 the Palestine government took a middle course, estimating the area of cultivable land outside the Negeb as 8,760,000 dunams.

Jewish experts, encouraged by the success of their reclama-

tion projects, are more optimistic. They believe that with systematic soil improvement and the restoration of certain mountainous regions, the total cultivable area of Palestine can
reach fourteen to sixteen million dunams. Their colonies established in some of the dismal areas demonstrate that they
have the courage and the ability to make wastelands sustain
them. In addition, Jewish experts maintain large areas could
be useful if reforested or turned into scientifically improved
grazing grounds.

On the strength of my own observations and of the vastly
increased economic possibilities under the proposed Jordan
Valley Authority, I accept the more optimistic point of view
on the total cultivable area of Palestine. I am further convinced that much more of Palestine can be irrigated than
even Mr. Fohs has estimated. As I have already pointed out,
without in the least disturbing the revered Sea of Galilee,
some of the waters of the Upper Jordan can be diverted to
the Valley of Esdraelon, while the excess flood waters of the
Yarkon can be utilized for the irrigation of the Negeb. Intensive farming on this irrigated area, supplemented by scientifically conducted grazing and improved dry farming, should
make possible a new agricultural population of about one
million above the present number of Jewish and Arab
farmers.

The extent of the arable area of Trans-Jordan is subject to
the same controversy as that of western Palestine. According
to a report of the High Commissioner for Palestine to the
League of Nations for the year 1936, the cultivable area of
Trans-Jordan is 4,600,000 dunams: cultivable land is defined
as "land which has been actually under cultivation at some
recent time." The Jewish Agency, however, in a memoran-

dum submitted to the Palestine Royal Commission in 1937, states that 4,100,000 dunams of additional land can be brought into cultivation without excessively costly reclamation work. I am convinced that a considerable part of Trans-Jordan's cultivable area can be irrigated if the JVA project is carried out.

The absorptive capacity of Palestine's agriculture can, I have attempted to show throughout this work, be greatly increased under scientifically controlled conditions. It is the agricultural population which serves as the base of every country's economic structure. In the United States farmers are a quarter of our total population and in normal times produce more food than we are able to consume. The industrialized countries of Europe have found that no more than a quarter of their population need be engaged in farming, for these countries are able to import additional food in exchange for their surplus of manufactured goods. Palestine, situated on the hub of three continents and in the immediate vicinity of great sea and air communication lines, can have easy access to foodstuffs and raw materials not produced in sufficient quantities in the country itself. The energy and ability shown by Jews in the development of Palestine's industry, as well as the increased industrial activity of Arabs, assure us that, in the better world order created by a democratic victory after the war, the industrial development of Palestine can readily be accelerated, making possible a large influx of needy but capable and industrious refugees. These industrial prospects will be further stimulated when the Jordan Valley Authority furnishes Palestine with a far greater quantity of cheap hydroelectric power than is available at the present time. The JVA will also undertake to enlarge even further the work now carried on at the Dead Sea in the extraction of the tremendous and very valuable mineral deposits which can serve as the foundation for important chemical industries.

Palestine's absorptive capacity can also be increased by full exploitation of its geographic position on the shores of the Mediterranean, with a direct outlet to the Indian Ocean through the Red Sea. Even before the war Palestine was one of the most profitable and attractive countries from the point of view of shipping interests. Jewish economic leaders in Palestine believe that after the end of the war they will be able to bring immigrants and freight in their own ships and ultimately make Palestine an important maritime center. Furthermore, the waters of the Mediterranean, and particularly of the Red Sea, are rich in fish; a scientifically developed fishing industry will be a valuable asset to Palestine's food supply. Maritime and fishing pursuits, as well as the commercial possibilities inherent in Palestine's geographic position, will enable many thousands of persons to support themselves in Palestine.

In estimating Palestine's capacity to absorb immigrants we must consider the distribution of wealth among them and the social structure they build. Colonists who rely on the exploitation of natives require large areas of land and cannot develop an intensive type of agriculture capable of forming the base for a dense population. Cheap native labor serves as an obstacle in the way of immigrants seeking work. Jewish settlement in Palestine is one of the very few instances in which European colonization has raised the standards of the native population. As we have seen, the Arabs of Palestine are not only more advanced and prosperous than those of the neighboring countries of Iraq and Syria, but have actually doubled their number during precisely the period which has brought hundreds of thousands of Jewish immigrants to Palestine. Moreover, by broad application of the principles of co-operation and conservation, the Jews of Palestine have created a social structure that supplies their own food with growing efficiency and produces industrial goods on an ever increasing scale.

I shall not attempt to estimate the final absorptive capacity of Palestine. That would be impossible, for the absorptive capacity of any country is a dynamic and expanding conception. It changes with the ability of the population to make the maximum use of its land, and to put its economy on a scientific and productive basis. It is clear, however, that there is ample proof of the assertion made in our chapter on the Jordan Valley Authority, that full utilization of the Jordan Valley depression for reclamation and power will in time make possible the absorption of at least four million Jewish refugees from Europe, in addition to the 1,800,000 Arabs and Jews already in Palestine and Trans-Jordan.

It is interesting to note that British scholars of the Palestine Exploration Fund who explored Palestine during the second half of the nineteenth century had a very high estimate of the country's capacity. That group of valiant explorers—General Charles Warren, General Sir Charles William Wilson, Colonel Claude Rainier Condor and Lieutenant, subsequently Field Marshal, Kitchener—were all enthusiastic about the possibilities of further development in Palestine. Sir Charles Warren expressed their common opinion in a book on Palestine issued as early as 1875, before even the earliest modern Jewish settlements were founded:

Give Palestine a good government and increase the commercial life of the people and they may increase tenfold and yet there is room. The soil is so rich, the climate so varied, that within ordinary limits it may be said that the more people it contains, the more it may accept. Its productiveness will increase in proportion to labor bestowed on the soil until a population of 15 million may be accommodated there.[1]

Warren had in mind the historic extent of Palestine which is much larger than the present mandated area. Applying his figures to present-day Palestine and Trans-Jordan, we may say that he and his colleagues appraised the country's absorptive capacity at twelve million. What has been done by Jewish settlers in the last six decades would have seemed to Warren and his colleagues merely confirmation of their vision.

On 14 per cent of the cultivated area and 6 per cent of the total area of Mandated Palestine, a people with faith and devotion born of long tradition has changed desolation into fertile fields, fruitful orchards and reforested slopes. Ancient cities have been rebuilt and the commerce on their streets quickened, long-unknown resources have been brought into the light of day and sent to the distant marts of the world. After the centuries of darkness which crushed the hopes of Palestine's miserable inhabitants, a new force has come into the land and made it live again. The possibility of a new day for the entire Near East is hidden in the fertile lands, the flourishing villages and cities, the co-operatives and the factories of Jewish Palestine.

If the forces of reclamation and progress Jewish settlers have introduced are permitted to continue, Palestine may

[1] *The Land of Promise*, pp. 5-6. London, 1875.

well be the leaven that will transform the other lands of the Near East. Once the great undeveloped resources of these countries are properly exploited, twenty to thirty million people may live decent and prosperous lives where a few million now struggle for a bare existence. Palestine can serve as the example, the demonstration, the lever, that will lift the entire Near East from its present desolate condition to a dignified place in a free world.

CHAPTER XV

ENGINEERED FEATURES OF THE PROPOSED JORDAN VALLEY AUTHORITY

. . . .

SINCE the first edition of this book, the irrigation and power features of the proposed Jordan Valley Authority have been engineered by the Commission on Palestine Surveys, of which James B. Hays, formerly of the engineering staff of the Tennessee Valley Authority, is chief engineer. During the winter of 1942-43 this commission carried out technical and engineering investigations into the possibilities of irrigation and hydroelectric power development in Palestine. First a thorough search was made for all maps, former surveys, reports, investigations, water-supply records and geologic data. Later a field study of several months' duration was made on the ground in Palestine to locate and examine sites for dams, canals, and powerhouses. This was done before the end of the war in Europe. The report was finally completed in February, 1946, and will be published in full in due time.

In this supplement I wish to summarize the essential features of the power and irrigation project as it has been engineered, passed on, and approved by a consulting board and by John L. Savage, the consulting engineer. The project has been found feasible and practical. Mr. Savage in his report to the State Department is more optimistic over the project than I was on the basis of my preliminary estimates. For he reports that the project would cost something less than 250 millions of dollars, and would pay out in fifty years at 3 per cent. I had not looked upon the original cost as a commercial venture but as a cost of war for the survival of a people, even as we do not look upon the costs of World Wars I and II as commercial enterprises. By rights the cost of the Jordan Valley Authority should be drawn from reparations from Germany in the Peace Treaty as restitution in small part of the eight billions of dollars' worth of Jewish property appropriated and confiscated in Germany and countries under German control during the war.

The purpose of the Jordan Valley project is to make full use of the unique resources of this remarkable area. To do so calls for treating the natural unity of the area, which partition of Palestine would violate. Palestine is so similar to southern California that what is possible in southern California is possible in Palestine on a proportionate basis. Like California, Palestine has more land suitable to irrigation than it has water for that purpose. Land development for intensive agriculture is limited by available waters. But its possibilities for hydoelectric power development are unusual.

The engineering plan calls for conservation and utilization of all surface and underground waters in or flowing into Palestine, including allowances for the Trans-Jordan side of the Jordan Valley, that could be developed economically. Investigations were also made for the use of surplus and wasted waters adjacent to Palestine that could be brought into the country within the limits of reasonable cost. Such use of waters for irrigation would call for the diversion of all waters from the Jordan River that now flow into and are lost by evaporation from the Dead Sea.

To replace the fresh waters of the Jordan River that are lost in the Dead Sea, the plan calls for diverting Mediterranean sea water through canals and tunnels to be constructed to and into the valley of the Jordan, for dropping this flow through nearly 1,300 feet of fall to the Dead Sea whose surface lies

[231]

nearly 1,300 feet below sea level in a hot furnacelike climate of 2 inches rainfall per year. This river of sea water about equivalent to the flow of the Jordan River would be made to create hydroelectric power by passing through turbines in two powerhouses.

The project is broken down into eight stages, for convenience in carrying out the work in consecutive stages to meet the needs, or in combining stages if desirable to speed up construction. It is necessary to maintain a proper sequence in the stages since certain works or provisions for one stage must be taken care of before a following stage can be undertaken. Seven of the stages provide water for irrigation and one for hydroelectric power. If waters from outside Palestine are obtained at a later time, two additional stages are possible, making ten in all, which could be completed in ten years' time. The main engineering report, however, does not include stages 9 and 10 but is concerned only with waters arising within Palestine.

The stages are summarized as follows:

1) *Stage One* starts with the development of underground waters in the coastal plains by drilling wells. This work would take place chiefly in the Emek Esdraelon and south coastal plain. This stage was set first because waters for irrigation could be made available in less than a year to irrigate about 175,000 acres of new land. A large amount of power will be required for this pumping. Diesel engines would be used in some cases and electric power in others. It is also proposed to build a storage dam on one of the main tributaries of the upper Jordan, which would be used in Stage Two. The reservoir being located at a comparatively high level would permit the generation of electric power during the summer months when large releases of irrigation water would give a heavy power output when most needed for pumping.

2) *Stage Two* includes the diversion of the summer flow of the several streams and springs forming the head waters of the Jordan River, the storage of the winter flow of one stream and the con-

veying of these waters through a "high line" gravity canal down the west side of the Jordan Valley. This canal would cross the divide to the Mediterranean slopes at the eastern end of Sahl-al Battauf. The areas to be irrigated would include a part of the upper Jordan (Huleh Basin), good lands in the Galilee Hills between the Emek and Tiberias and a large part of the Emek Esdraelon. This stage would take away a part of the normal flow of the Jordan River and would therefore affect the power output of the electric plant at Naharayim.

3) *Stage Three.* In order to be prepared for further diversions of water from the Jordan, the third stage calls for the diversion of the flow of the Yarmuk River into Lake Tiberias (Sea of Galilee). By the time all of the upper Jordan River waters have been diverted, there would not be sufficient water flowing into Lake Tiberias to provide for annual evaporation. Water from another source must be provided. This is done by diverting Yarmuk River water into Lake Tiberias. The lake serves also as storage of waters due to the diversion of the upper Jordan. The Jordan and the Yarmuk rivers have practically equal annual discharges where they join. The Jordan takes out of Lake Tiberias about 85 per cent of the inflow at the upper end. All the rest is lost in evaporation including the inflow of other streams and springs.

The waters of the Yarmuk, when put under control, would be used to irrigate lands on both sides of the Jordan Valley: half to the Trans-Jordan side and half to the Palestine side. At present these waters are not available to the Jordan Valley lands, but have been assigned to the Palestine Electric Corporation for the generation of electric power. Irrigation would produce benefits far more valuable to the people and government than is now possible from the comparatively small amount of electric power produced. Trans-Jordan's share plus the conserved waters of all the *wadis* or *strems* originating in the mountains of that area and flowing into the Jordan Valley would be sufficient to irrigate all the suitable lands on the east side of the Jordan between the Yarmuk and the Dead Sea.

The remainder of the Yarmuk River water after evaporation in Lake Tiberias would be used in Palestine south of the lake and in the lower areas of the Jolud Valley on the west bank of the Jordan.

4) *Stage Four* includes the completion of the Mediterranean–Dead Sea power system. The demand for electric power would be increasing constantly for growing industries and the following stages call for the diversion of all the Jordan River water that can be economically done. The Jordan River contributes about one-half of the average annual supply to the Dead Sea, where it is lost by evaporation. This proposed power scheme will replace the fresh waters with Mediterranean sea water to maintain the level of the Dead Sea. Otherwise it is estimated that the Dead Sea would shrink to about one-half its present surface area and about 100 to 150 feet lower in elevation.

The flow of salt water to be let in from the Mediterranean Sea would average more than 1,000 cubic feet per second. This sizable stream would be conducted through lined canals and penstocks and would be made to pass through turbines of two great hydroelectric power plants before it reached the level of the Dead Sea. The hydroelectric power so developed would equal 560 million kilowatt hours per year. This power in turn would be supplemented by power plants on fresh-water canals and at reservoirs of the irrigation system by another 100 million kilowatt hours making a total of 660 million kilowatt hours expected from power developments of eight stages of the project.

5) *Stage Five.* In Stage Two, the summer waters of the upper Jordan, plus the year round flow of one tributary, have been made use of. But the heaviest run-off takes place in winter. Having provided for maintaining levels of Lake Tiberias and of the Dead Sea it is now possible to divert or impound the winter flow or balance of the upper Jordan supply. Fairly large reservoir capacities are required to include "carry-over" storage from year to year so as to balance out variable annual flow. Other than the storage proposed in a previous stage, there are no suitable storage sites on the upper Jordan River or on its tributaries. It is, therefore, planned to convey such waters to a large reservoir in the Sahl-al Battauf, in a flat valley north of Nazareth, with a narrow outlet to the west. This proposed reservoir site is located on the route of the canal mentioned in Stage Two. Waters thus recovered would be used

[234]

to irrigate the remainder of the Emek Esdraelon and leave some
to be carried by syphon across the Qishon River narrows and
canals to the coastal plain south of Mount Carmel. Since most of
the coastal plain area in this section is comparatively rich in underground waters, the waters in this canal would be carried on
south until it is used up for irrigation and domestic use.

6) *Stage Six* includes the drainage and reclamation of the
Huleh. With the floods of the Upper Jordan under control the
next work would consist of drainage and reclamation of the Huleh
Lake and swamps. The marshes, when properly drained, would
open a highly fertile area for intensive farming. This, together
with the drainage of the shallow lake, would greatly reduce the
serious malarial infestation of this area. The enormous evaporation from Lake Huleh and the marshes, when controlled, is more
than sufficient to supply all the irrigation water needed for the
farm land, and a surplus that could be pumped into the main
"high line" canal leading to the Battauf Reservoir and to the south
coastal plain.

7) *Stage Seven* includes the storage of waters along the western slopes of the Jordan Valley. Having developed the upper
Jordan waters and the Yarmuk River, there still remains a flow
into the Jordan River not yet accounted for. This is mostly from
the west side, between Lake Tiberias and the Dead Sea, plus return flow from irrigation of Jordan Valley lands in Trans-Jordan
and Palestine from Lake Tiberias to the lower end of the Beisan
plain.

A dam across the Jordan River where the Wadi el Malik en-
ters from the west would be built to feed a canal taking water
to irrigate a large part of the lands of the Jericho plain. Additional water for this area—enough to supply all the remaining
good lands—could be obtained by storage and conservation of the
flow of several wadis flowing across the plain to the Jordan River.
Most of the soils in this area are impregnated with salts in varying amounts. The drainage and underground conditions are such,
however, that by application of large quantities of irrigation water
at first, these soils may be leached of noxious salts suitable for
growing specialty crops. Some of the worst soils of the Jordan

Valley have been leached in this manner successfully and give promise for the method on other lands less salty.

The warm winter climate of the deep Jordan Valley favors the growing of out-of-season vegetables, as does the Imperial Valley of California. The hot summer climate is not worse than that of the Imperial Valley and is not as bad as it might be because of light breezes that blow daily up and down the valley.

8) *Stage Eight* includes the collection and storage of all flood waters from the coastal valleys or wadis of Palestine, together with other surplus and winter waters of springs in or contiguous to the project area.

These flood waters are erratic, and the quantities that may be recovered from year to year will vary from small amounts to several times the average flow. All depends on the amount and distribution of rainfall. Numerous light rains spread out through the rainy season do not produce as much run-off as the same amount of rainfall that comes in a few heavy downpours. Regular measurement of the flow of many of the wadies has been started.

Stage Eight has purposely been left to the last in the general scheme to allow time for investigations of the flows of the coastal wadis that may be counted on for designs of structures. It would, however, be possible to go ahead with construction of any part of Stage Eight that it is justified and to use it alone or in combination with any other stage if a satisfactory estimate of the stream flow can be made.

Under the proposed plan all preceding stages have been balanced to make available an average flow each year in light or heavy rainy seasons. Stage Eight uses a more erratic water supply and has less favorable storage conditions, but not impossible for "carry-over." If, however, this stage is combined with preceding stages and all are placed under the same operating authority, as contemplated in the Jordan Valley Authority, the water supply can be shifted so that a large area will suffer only a slight loss rather than a small area suffer heavy damage in a season of light rainfall.

The general plan is based on supplying lands nearest the

source of water supply first and on expanding the irrigated area as additional water sources are developed. Following such a program, step-by-step development leaves the irrigation of the Negev until the last. The plan, nevertheless, has a certain amount of flexibility that would permit the earlier irrigation of parts of the Negev, but on somewhat higher unit costs.

The amount of water required annually per unit of land to grow crops varies throughout the country due to climatic conditions. The northern areas with heavier rainfall require the least water for irrigation. The Negev will require twice as much and the Lower Jordan Valley three times as much as needed in the North. Estimates as outlined are based on irrigation by flooding and furrows. Sprinklers for overhead irrigation as now used in many parts of southern California, have been adopted by many of the farm settlements, and more will be installed when materials become available. For overhead irrigation has proved more economical as a result of many comparative measurements. These records also show a net saving in labor and approximately a one-third saving in water. This saving in water is important for it permits the expansion of the proposed irrigation project from 604,000 acres (2,416,000 dunams) of new watered land to about 750,000 acres (3,000,000 dunams) in addition to the 100,000 acres now irrigated or a total of over 800,000 acres suitable to two and three crops annually.

The cost of the proposed scheme is not excessive as compared to irrigation projects of southern California. On the basis of the last-reported reliable costs available at the time, it is estimated water delivered to the farm will cost on the average 2¼ *mils* (P) per cubic meter. At present farmers in Palestine are paying 3 to 4½ *mils* (P) per cubic meter for irrigation water.

The average amount of water for the proposed project required annually to grow crops is 2.66 acre feet per acre—that is, a depth of 2.66 feet over each acre in addition to rain fall. The annual cost of water including the repayment of capital investment amortized over a period of fifty years at 3 per cent would be $22.40 per acre or $8.42 per acre foot. In southern California, water costs average about $9.11 per acre foot, according to the California Division of Water Resources. California farmers use somewhat less water per acre—from 1.75 acre feet in the Los Angeles basin to 5.00 acre feet in the Imperial Valley. Water use and costs in the two areas are not out of line.

New lands require more water than soils that have been irrigated and properly farmed for some time due to accumulation of humus in the soil. It is expected that Palestine farmers could reduce the amount of water to the same degree in the course of time.

The total amount of power to be generated in the system by making use of all waters in or flowing into Palestine amounts to about 660 million kilowatt hours per annum with an installed capacity of 163,700 kilowatts. The additional power that would be generated by stages 9 and 10 of further irrigation developments from importation of outside surplus waters would be about 400 million kilowatt hours per year with an installed capacity of 68,000 kilowatts. Full development of possibilities for power in Palestine would total over one billion kilowatt hours per annum or about one-fifth the power developed at Boulder Dam.

A great irrigation and power project such as that which the unique features of the Jordan Valley and the maritime slopes of Palestine suggest may offer a constructive approach to the political conflict that now burdens the Holy Land. It will give work and opportunity to all Jewish refugees from Europe who wish to go to Palestine—without displacing any Arab or Christian now in Palestine. On the contrary it will improve the lot of all and will serve as a demonstration as to how the old and wasted lands of the Middle East may be restored to a condition worthy of their glorious past.

INDEX

Aaronson, Aaron, 53-54
Absorptive capacity
  factors in, 219-221, 227
  of Palestine, 218-229
Agriculture (*see also* Terrace agriculture)
  ancient, 63-64, 68-69
  Bedouin, 193-194
  co-operative, 126-128
  development of, 86-90, 93-100, 102, 198
Air traffic, 120
Anti-Semitism, 6, 7-10
Arab invasions, early, 70-71, 73-74
Arab riots, 3, 4, 132, 153, 166, 216
Arabs
  advantages to, of Jewish immigration, 131, 132, 149, 151-167, 226-227
  attitude of, 80
  benefits of JVA to, 178-179
  collaboration of, with Jews, 164-165
  economy of, 149-154
  emigration of, 154-155, 216
  employment of, in Jewish enterprises, 162-163
  opposition of, to Jewish immigration, 161-167
Armunk, Juster, cited, 57
Assimilation of Jews under Arab rule, 166
Atlantic Charter, 18
Avi-Yonah, M., cited, 57

Balfour Declaration, 8
Balsam
  ancient production of, 62
  disappearance of, 78

Baron, Salo, cited, 58
Bedouins, 191-194, 199-200
Birth rate, Arab, 157
British Royal Commission, quoted, 139, 151-152, 153-154
Buechler, Adolph, cited, 58
Building industry
  in ancient times, 65
  modern, 110-112
Building materials, 56
Burckhardt, John Lewis, 202

California, similarities of Palestine to, 39-41, 51-52
Carnet, John
  cited, 79
  quoted, 78
Cereal crops, 99-100
Chemical industry, ancient, 63
China, erosion in, 82
Cisterns, use of, 41, 45, 185
Citrus crops, 88, 94-97
Civilization
  debt of, to Near East, 12-15
  and food, 19-20
  pre-war, break-up of, 15-16
Climate, 30, 35-39
  of the Dead Sea Valley, 33, 146
  of Galilee, 49
  similarity of, to that of California, 39-41
  theory of pulsations in, 81-83, 190
  of Trans-Jordan, 210-211
Collective settlements, 127-128
Colonization of Palestine, modern, 14-15, 24, 85-103
Conder, Col. C. R., cited, 77, 227
Conservation
  human, 18-19

239